37. Colloquium der Gesellschaft für Biologische Chemie
10.–12. April 1986 in Mosbach/Baden

Cell Cycle and Oncogenes

Edited by
Widmar Tanner and Dieter Gallwitz

With 64 Figures

Springer-Verlag
Berlin Heidelberg New York London Paris Tokyo

Professor Dr. WIDMAR TANNER, Naturwissenschaftliche Fakultät III,
Biologie und Vorklinische Medizin, Universitätsstraße 31, 8400 Regensburg, FRG

Professor Dr. DIETER GALLWITZ, Physiologisch-Chemisches Institut I,
Universität Marburg, Emil-Mannkopff-Straße 1, 3550 Marburg, FRG

ISBN-13:978-3-642-71688-1 e-ISBN-13:978-3-642-71686-7
DOI: 10.1007/978-3-642-71686-7

Library of Congress Cataloging in Publication Data. Gesellschaft für Biologische Chemie. Colloquium (37th: 1986: Mosbach, Baden-Württemberg, Germany) Cell cycle and oncogenes. Includes bibliographies and index. 1. Oncogenes–Congresses. 2. Cell cycle–Congresses. 3. Microbial cell cycle–Congresses. I. Tanner, W. (Widmar), 1938– . II. Gallwitz, Dieter. III. Title. [DNLM: 1. Cell Cycle–congresses. 2. Cell Transformation, Neoplastic–congresses. W3 GE382R 37th 1986c/QZ 202 G389 1986c] RC268.42.G47 1986 616.99'4071 86-29849

Softcover reprint of the hardcover 1st edition 1986

2131/3130-543210

Contents

I. The Eukaryotic Cell Cycle and Its Regulation

II. Cellular Oncogenes and Their Relation to the Cell Cycle

III. Growth Factors: Primary Effects and Relation to Oncogenes

Contributors

You will find the addresses at the beginning of the respective contribution

I. The Eukaryotic Cell Cycle and Its Regulation

Cell Cycle Genes as Potential Oncogenes

R. Baserga, L. Kaczmarek, B. Calabretta, R. Battini, and S. Ferrari[1]

Approaches to the Identification of Genes Controlling Cell Proliferation

The control of cellular proliferation in animal cells is regulated by growth factors and by the cellular genes and gene products that respond to the growth factors. The search for genes and gene products that control cell reproduction began more than 20 years ago with the pioneering experiments of Lieberman et al. (1963), and of Baserga et al. (1965). These early experiments, based on the use of inhibitory drugs like actinomycin D, have been previously reviewed (Baserga 1978, 1985). More recently, a formal demonstration was obtained that RNA polymerase II-directed transcripts are necessary for the entry of quiescent cells into S phase (Rossini et al. 1980, Baserga et al. 1982). It is therefore reasonable to search for genes transcribed by RNA polymerase II that control cell proliferation in animal cells.

There are essentially three different approaches to the identification of genes controlling cell reproduction in animal cells, namely (1) the identification of protooncogenes, i.e., the cellular equivalents of retroviral transforming genes; (2) The isolation and molecular cloning of the genes that can correct defects in temperature-sensitive mutants of the cell cycle; and (3) a more systematic analysis, i.e., the identification of sequences preferentially expressed in a specific phase of the cycle by differential screening of cDNA libraries.

Of these three approaches, the isolation of genes that can correct temperature-sensitive defects of the cell cycle has been so far much more successful in yeasts than in mammalian cells and will be treated elsewhere in this Colloquium. The protooncogenes will also be discussed by other colleagues. I will therefore describe the remaining approach as well as the functional criteria that should be used to determine the possible role played by a gene in the control of cell proliferation.

Screening of cDNA Libraries

The screening of a cDNA library offers a systematic approach to the identification of cell cycle genes. The strategy that is used for this approach is straightforward, and it has been used for the identification and the molecular cloning of a variety of genes whose expression varies in different physiological conditions. For example, in our laboratory, we built a cDNA library from poly(A)$^+$ RNA isolated from ts13 cells that had been stimulated to proliferate with serum for 6 h (G_1 cells). The cDNA's were cloned in pBR322 (Hirschhorn et al. 1984) and replica plates of the library were made. The replica plates were screened with single stranded cDNA made from either of two sources, poly(A)$^+$ RNA of G_0 cells, and poly(A)$^+$ RNA of G_1 cells (i.e., other

[1]Department of Pathology and Fels Research Institute, Temple University Medical School, Philadelphia, PA 19140, USA

37. Colloquium Mosbach 1986
Cell Cycle and Oncogenes

cells stimulated for 6 h with serum), in order to identify cDNA clones that hybridized only to the single-stranded cDNA of G_1 cells and did not hybridize to the single-stranded cDNA of G_0 cells. When the screening is repeated at least three times, one ends up with a number of clones that, indeed, can be said to be expressed only in a specific phase of the cell cycle, in the example given above only in cells stimulated to proliferate. This strategy is the same as that followed by other laboratories (Linzer and Nathans 1983, Cochran et al. 1983, Arya et al. 1984, Matrisian et al. 1985a,b, Edwards et al. 1985, Lau and Nathans 1985), and it invariably results in the isolation of some interesting genes specifically expressed in a phase of the cell cycle. It should be stated at this point that the great majority of genes are not expressed in a cell cycle-dependent manner. In the laboratories mentioned above, a common finding was that only between 0.3 and 1% of the genes were expressed in a cell cycle dependent manner.

Cell Cycle-Dependent Genes

Using this procedure we have isolated a number of genes that are preferentially expressed when quiescent cells are stimulated to proliferate and that were described in the paper by Hirschhorn et al. (1984). The pitfalls of this approach are clear. First of all, although this approach identifies genes that are expressed in a cell cycle-dependent manner, the fact that they are expressed in a cell cycle-dependent manner does not mean that they regulate cell proliferation. They may regulate the cell cycle, or they may be regulated by the cell cycle. This is a fundamental difference that one should keep in mind and it is often ignored, especially with protooncogenes. Another problem with this approach is that cDNA libraries are usually made from poly$(A)^+$ RNA and some of the interesting RNA's may be poly$(A)^-$. Finally, there is no *a priori* reason why a gene regulating cell proliferation should be expressed, in terms of RNA levels, in a cell cycle-dependent manner. It may be possible that it is expressed in a non-cell cycle-dependent manner throughout the cell cycle, and yet regulate cell proliferation by translational controls. A spectacular instance of such a behavior is the cdc 2 gene of *S. pombe*, the "start" gene of fission yeasts, which is *not* cell cycle-dependent.

However, the importance of cell cycle-dependent genes in cell proliferation is strengthened by the finding that several bona fide oncogenes are expressed in a cell cycle-dependent manner. For instance, c-*myc* is induced by platelet-derived growth factor in quiescent mouse 3T3 cells, and by lipopolysaccharide or concanavalin A in mouse G_0 lymphocytes (Kelly et al. 1983, Campisi et al. 1984) and by PHA in human lymphocytes (Kaczmarek et al. 1985b), and so is c-myb (Torelli et al. 1985). p53 is inducible in quiescent 3T3 cells by serum (Reich and Levine 1984). More recently p53 was shown to be an oncogene that can replace c-*myc* in transformation experiments (Eliyahu et al. 1984). c-*ras* is also expressed in a cell cycle-dependent manner (Campisi et al. 1984), and finally c-fos has been shown by several investigators to be induced by stimulation of G_0 cells with either serum or platelet-derived growth factor or EGF (Greenberg and Ziff 1984, Kruijer et al. 1984, Muller et al. 1984). The fact that these oncogenes are expressed in a cell cycle-dependent manner by itself does not mean that they regulate the cell cycle, as already discussed above. However, the fact remains that these oncogenes, when appropriately modified by retroviral vectors, are known to transform cells and therefore they must have some influence on growth regulation. We can now turn this statement around and say that if oncogenes are expressed in a cell cycle-dependent manner, especially during the transition from G_0 to G_1, then one can look at cell cycle-dependent genes, especially genes induced in G_0 cells as possible candidates for

Table 1. Summary of biological characteristics of selected cell cycle genes

Cell gene	(a) type	(b) species	(c) CX	(d) ts	(e) tumors	(f) Adeno	(g) m.i.	(h) m.i.inhib.	(i) senescent	(j) PHA
4F1	+	+	+	+	+	–	?	?	+	–
2F1	+	+	+	+	+	+	+	?	+	+
2A9	+	+	+	+	+	–	+	?	+	–
c-*myc*	+	+	+	+	+	–	+	–	+	–
p53	+	+	+	?	+	+	+	+	+	–
H3	+	+	+	–	–	+	–	+	+	–

a) Inducibility by different mitogens in different cell types; b) Inducibility by mitogens in cells of different species; c) Inducibility by mitogens in the presence of cycloheximide; d) Inducibility in G^1-ts mutants at the restrictive temperature; e) Overexpressed in transformed cells and/or human neoplasias; f) Inducible by adenovirus infection; g) Microinjection of gene (or gene product) stimulates DNA synthesis; h) Microinjection of monoclonal antibody inhibits serum stimulated DNA synthesis; i) Inducible in senescent human diploid fibroblasts; j) Induced by PHA in purified T lymphocytes.
References to this table are found in the text.

potential new oncogenes. In other words, other G_1 genes, whose expression is induced in the G_0 to S transition may, in actuality, be genes potentially capable of altering growth regulation. The fact that they have not been picked up by retroviruses should not be looked upon too disfavorably. After all, the p53 gene can immortalize cells and transform them, and yet it has never been identified in any of the transforming retroviruses.

As already stated above, several of the cDNA clones that have been isolated by this approach have always proved to be interesting. For instance, one isolated by Linzer and Nathans (1983) turns out to have 70% homology with another growth factor, prolactin, and is identical to a mitogen regulated protein (Parfett et al. 1985). Three other genes isolated by Hirschhorn et al. (1984), specifically 2F1, 4F1, and 2A9, were also found to be considerably interesting. They are expressed not only in hamster and mouse cells stimulated by serum, but also in human lymphocytes stimulated by phytohemagglutinin (Kaczmarek et al. 1985b). Even more interesting, they are overexpressed, like oncogenes, in certain forms of human leukemias (Calabretta et al. 1985, 1986). They are also induced by serum even in the presence of an inhibitor of protein synthesis like cycloheximide (Rittling et al. 1985) and they are induced even in G_1 ts mutants stimulated at the restrictive temperature. Some of these data are summarized in Table 1, where these 3 cDNA clones are compared to the S phase gene, histone H3, and to 2 oncogenes, c-*myc* and p53. Notice that all of them are expressed in different cell types, and in cells of different species, they are still induced in the presence of cycloheximide and in G_1 ts mutants of the cell cycle, stimulated at the restrictive temperature. Notice also that the three G_1 cell cycles genes are overexpressed in certain human tumors, just like the oncogenes. An interesting side observation is that adenovirus preferentially stimulates late G_1 and S phase genes and has very little effect on the early or mid G_1 genes (Liu et al. 1985).

Function of Oncogenes or Cell Cycle Genes

We will divide this section into two parts. In the first part we shall discuss the methods that have been used to establish the role of a gene in the control of cell proliferation. These are functional assays, that have only a modest value for screening purposes but that can and have been used successfully to determine whether or not a gene and its product play a role in cell cycle progression. The two methods are: (1) Microinjection of genes or gene products into quiescent animal cells, to determine whether they are capable of stimulating cellular DNA synthesis: (2) Microinjection into quiescent animal cells of monoclonal antibodies to gene products, with subsequent stimulation of the microinjected cells, in order to determine whether the microinjected monoclonal antibodies may inhibit serum stimulated DNA synthesis. A variant of this approach is the microinjection of antisense RNA. In the second part, we shall discuss how DNA homologies can be used to identify possible functions.

Microinjection of Genes or Gene Products

The manual microinjection of a gene, or a gene product into quiescent animal cells is the most direct approach to the identification of genes that control the G_0-S transition. Obviously if a microinjected gene or gene product can stimulate cellular DNA synthesis in G_0 cells they must be considered as playing a major role in the control of cell proliferation, similarly to the role of the SV40 T-antigen coding gene or the "start" gene of yeasts. There are, however, some qualifications.

It is well established that transforming genes from oncogenic DNA viruses can stimulate cellular DNA synthesis in G_0 cells (for a review see Baserga 1985). More recently it was demonstrated that the Harvey v-*ras* gene, when microinjected into G_0 cells, can stimulate cellular DNA synthesis (Hyland et al. 1985). The same results can be obtained by microinjecting the activated c-*ras* protein (Feramisco et al. 1984, Stacey and Kung 1984). Thus one can say that the v-*ras* protein, or the activated c-*ras* protein are capable to initiate cell cycle progression. Surprisingly the microinjection of v-*myc*, which is also an oncogene, failed to induce cellular DNA synthesis in quiescent 3T3 cells (Hyland et al. 1985). The explanation came in subsequent experiments (Kaczmarek et al. 1985a), in which a c-*myc* protein (produced in bacteria by recombinant DNA technology) was microinjected into quiescent cells. Microinjection of c-*myc* protein into G_0 Swiss 3T3 cells failed to stimulate cellular DNA synthesis unless platelet-poor plasma was added after microinjection. Platelet-poor plasma (PPP) does not contain platelet-derived growth factor (PDGF) and is the progression factor in the competence progression theory of the G_0 to S transition that has been proposed by several investigators (for a review see Scher et al. 1979). By itself, PPP does not stimulate cellular DNA synthesis, although it contains other growth factors that are important for cell cycle progression, like insulin-like growth factor I (IGF-1). It seems, therefore, that c-*myc*, when microinjected into G_0 cells, to use the competence-progression language, makes the cells competent but that other growth factors are needed before the cell can enter into the S phase. Therefore, to identify genes or gene products that can initiate cell cycle progression microinjection has to be done under different conditions, i.e., in the presence or absence of other growth factors. Similar results have been obtained with the microinjection of the molecularly cloned cDNA or gene for the p53 protein (Kaczmarek et al. 1986).

Table 2. Homologies between cell cycle genes, oncogenes, growth factors and receptors for growth factors

a) PDGF and v-*sis* (Doolittle et al. 1983)

b) EGF-R and v-*erb*B (Downward et al. 1984)

c) Transferrin and B-*lym* (Diamond et al. 1983)

d) CSF-1 R and v-*fms* (Sherr et al. 1985)

e) Proliferin and prolactin (Linzer and Nathans 1984)

f) Insulin-R and v-*ros* (Ebina et al. 1985)

Clearly, although, this approach can unequivocally establish the role of a gene in cell cycle progression, it can hardly be used as a screening procedure for identifying possible new genes involved in the regulation of cell proliferation.

Microinjection of Monoclonal Antibodies to Gene Products

In this approach which is, essentially, the reverse of the first one, a population of G_0 cells are microinjected with monoclonal antibodies to a certain protein and then stimulated with serum, or with other growth factors. If a particular protein plays a major role in cell cycle progression the microinjection of a monoclonal antibody that neutralizes it should result in the inhibition of cell cycle progression. This approach was first pioneered by Mercer et al. (1982) with a monoclonal antibody to the p53 protein, a transformation-related protein that is often found in higher amounts in transformed cells than in their normal untransformed counterparts (Linzer et al. 1979, Crawford et al. 1981, Lane and Crawford 1979, Linzer and Levine 1979, Ruscetti and Scolnick 1983). The synthesis of the p53 protein is markedly increased in lymphocytes stimulated by concanavalin A (Milner and Milner 1981) or by phytohemagglutinin (Mercer and Baserga 1985) and the expression of the p53 mRNA is also increased when G_0 cells are stimulated to proliferate (Reich and Levine 1984).

The pitfalls in this approach include all the pitfalls of the previous approach, plus the fact that the amount of monoclonal antibody that can be microinjected is limited. Apart from the uncertainties of the approach, its value for identifying new genes that may be involved in cell proliferation is limited. The use of anti-sense RNA instead of monoclonal antibodies adds sophistication to the approach, but not improvement in screening capacity.

Overlapping of Growth-Regulated Genes and Protooncogenes

We can also look at functions of cell cycle genes by determining the extent of overlapping between cell cycle-dependent genes, oncogenes, growth factors and receptors for growth factors. There are three lines of evidence indicating that there is a considerable amount of overlapping among these four sets of genes. These lines of evidence include (1) homologies in the DNA sequences, (2) cell cycle dependence of expression, and (3) overexpression of cell cycle-dependent genes in transformed cells. These three lines of evidence will be considered separately.

The DNA homologies between cell cycle-dependent genes, oncogenes, growth factors and receptors for growth factors are summarized in Table 2.

The two best-known ones are, of course, the homology between PDGF and v-*sis* (Doolittle et al. 1983), and between the receptor for EGF and v-*erbB* (Downward et al. 1984). However, in the past few years several other homologies among these sets of genes have been discovered. There is homology between transferrin, a growth factor, and the cellular oncogene B-*lym* (Diamond et al. 1983), and between the receptor for colony stimulating factor CSF-1, and v-*fms* (Sherr et al. 1985). As already mentioned, Linzer and Nathans (1984) showed a 70% homology between prolactin and a cell cycle-dependent gene (proliferin), which they isolated from serum stimulated mouse 3T3 cells. Other homologies include: an homology between the receptor for insulin, a growth factor, and the transforming gene, v-*ros* (Ebina et al. 1985). To this we can add the homologies between the start gene of yeasts and several vertebrate oncogenes (Lorincz and Reed 1984), and between the *ras* protooncogene and genes that control the growth of yeast cells (Powers et al. 1984). Clearly these homologies are more than casual. An interesting case of DNA homology, that of the 2A9 clone, will be illustrated below.

The second line of evidence is the cell-cycle dependence of expression of certain protooncogenes discussed above. The third line of evidence is the overexpression of certain cell cycle genes in transformed cells. Overexpression of oncogenes is an accepted mechanism for the activation of protooncogenes. While in some cases the transforming action of protooncogenes is due to mutations, or to insertion mutagenesis, in some other cases the transforming effect has been obtained simply by the overexpression of the normal cellular oncogene. There are instances in which the overexpression of cell cycle-dependent genes is associated with the characteristics of transformation. Apart from the case of the p53 gene, certain cell cycle-dependent genes are overexpressed in human leukemias (Calabretta et al. 1985, 1986). Three of these genes, 2F1, 4F1, and 2A9 are genes that were isolated as cDNA clones by differential screening of a cDNA library from ts 13 hamster cells. These three genes were shown to display a marked increase in the levels of expression in human acute and myeloid leukemias and in acute lymphoblastic leukemia. There was no increase in the levels of expression of these cell cycle dependent genes in chronic lymphocytic leukemia (Calabretta et al. 1985). Not all cell cycle dependent genes are overexpressed in these forms of leukemia. For instance, the two genes isolated by Cochran et al. (1983) from mouse 3T3 cells, JE-3 and KCl, although induced in peripheral blood mononuclear lymphocytes stimulated by PHA were not overexpressed in the cases of leukemia that were studied by Calabretta et al. (1985).

The 2A9 Genes

We have isolated from an Okayama-Berg library (Okayama and Berg 1983) the human cDNA corresponding to the Syrian hamster cDNA clone that was originally isolated by Hirschhorn et al. (1984) and called 2A9. In terms of steady-state RNA Levels, the gene corresponding to the 2A9 cDNA is not detectable in G_0 cells, increases after serum stimulation, and remains high in cycling cells. In this respect it behaves like c-*myc* (Hann et al. 1985, Thompson et al. 1984). 2A9 is inducible in 3T3 cells by PDGF and EGF but not by platelet poor plasma or insulin.

2A9 is inducible in a variety of cells of different types and from different species, but is not detectable in human peripheral blood mononuclear cells, either before or after mitogen stimulation (Calabretta et al. 1985). It is detectable in human bone marrow cells and in cells from acute and chronic and myeloid leukemias (Calabretta et al. 1985, 1986). Its expression is deregulated in some forms of human acute myeloid leukemia (Calabretta et al. 1986).

The sequence of 2A9 reveals a 55% homology with the coding sequence of the β-subunit of the S-100 protein. This protein was originally described as a brain-specific protein (Moore 1965, Isobe et al. 1981), but it has been more recently described in other cell types. The S-100 protein has been detected by immunofluorescence in several human tumors, especially malignant melanoma and tumors of neuroendocrine origin (Stefansson et al. 1982, Kahn et al. 1983), and is synthesized in the G_1 phase of cultured glioma cells (Fan 1982). The S-100 protein is a calcium-binding protein and has the two characteristics sequences that have been proposed as the calcium-binding sites and that are partially homologous to similar sequences in other calcium-binding proteins, like the intestinal calcium-binding protein, parvalbumin, troponin C and calmodulin (Cheung 1980). The 2A9 clone has two sequences that are strongly homologous to the calcium-binding sequences of the S-100 protein and of another protein, the vitamin D-dependent calcium-binding protein of intestinal origin (for a review, see Van Eldik et al. 1982). We find this homology between 2A9 and other calcium-modulated proteins as particularly intriguing, because of the well known role of calcium in cell proliferation (Hazelton et al. 1979).

There is another protein that is homologous to the S-100 protein and that is the small subunit of a protein complex that is a substrate for viral tyrosine-specific protein kinases and is called p11 by Gerke and Weber (1985) and p10 by Glenney and Tack (1985). Because tyrosine kinases are also involved in the response to growth factors (Cooper et al. 1982, Carpenter et al. 1982), these homologies can easily lead to speculations which, for the moment, we prefer to leave untouched except to mention that 2A9 may be the missing link between growth factors and cell cycle progression. Also on the speculative side is another role of calcium, i.e., the calcium mobilization that occurs when protein kinase C is activated (Kaibuchi et al. 1985, and review by Nishizuka 1984), as it occurs, for instance, with the transforming gene of Rous sarcoma virus (Sugimoto et al. 1984), growth factors like Interleukin-2 (Farrar and Anderson 1985), or EGF (Sawyer and Cohen 1981, Raben and Cunningham 1985) and polyoma middle T (Whitman et al. 1985).

Conclusions

We can now come to the conclusions, which I would like to summarize as follows: (1) there are cell proliferation genes, which include growth factors, receptors for growth factors, and regulatory proteins like those necessary for the DNA synthesizing machinery or other functions like Ca^{2+} binding; (2) protooncogenes and cell cycle-dependent genes, growth factors and receptors for growth factors are overlapping subsets of the larger set of cell proliferation genes; and (3) as a corollary, any alteration in any of these cell proliferation genes could lead to altered growth regulation.

References

Arya SK, Wong-Staal F, Gallo RC (1984) Transcriptional regulation of a tumor promoter and mitogen-inducible gene in human lymphocytes. Mol Cell Biol 4:2540-2542

Baserga R (1976) Multiplication and division in mammalian cells. Dekker, New York

Baserga R (1985) The biology of cell reproduction. Harvard Univ Press, Cambridge MA

Baserga R, Estensen RD, Petersen RO (1965) Inhibition of DNA synthesis in Ehrlich ascites cells by Actinomycin D II. The presynthetic block in the cell cycle. Proc Natl Acad Sci USA 54:1141-1148

Baserga R, Waechter DE, Soprano KJ, Galanti N (1982) Molecular biology of cell division. Ann NY Acad Sci 397:110-120

Calabretta B, Kaczmarek L, Mars W, Ochoa D, Gibson CW, Hirschhorn RR, Baserga R (1985) Cell-cycle specific genes differentially expressed in human leukemias. Proc Natl Acad Sci USA 82:4463-4467

Calabretta B, Venturelli D, Kaczmarek L, Narni F, Talpaz M, Anderson B, Beran M, Baserga R (1986) Altered expression of G_1 specific genes in human malignant myeloid cells. Proc Natl Acad Sci USA 83:1495-1498

Campisi J, Gray HE, Pardee AB, Dean M, Sonenshein GE (1984) Cell-cycle control of c-myc but not c-ras expression is lost following chemical transformation. Cell 36:241-247

Carpenter DE, Stoschek CM, Soderquist AM (1982) Epidermal growth factor. Ann NY Acad Sci 397:11-17

Cheung WY (1980) Calmodulin plays a pivotal role in cellular regulation. Science 207:19-27

Cochran BH, Reffel AC, Stiles CD (1983) Molecular cloning of gene sequences regulated by platelet-derived growth factor. Cell 33:939-947

Cooper JA, Bowen-Pope DF, Raines E, Ross R, Hunter T (1982) Similar effects of platelet-derived growth factors and epidermal growth factor on the phosphorylation of tyrosine in cellular proteins. Cell 31:263-273

Crawford LV, Pim DC, Gurney EG, Goodfellow P, Taylor-Papadimitriou J (1981) Detection of a common feature in several human tumor cell lines; a 53,000 dalton protein. Proc Natl Acad Sci USA 78:41-45

Diamond A, Cooper GM, Ritz J, Lane MA (1983) Identification and molecular cloning of the human B lym transforming gene activated in Burkitt's lymphomas. Nature (London) 305:112-116

Doolittle RF, Hunkapiller MW, Hood LE, Devare SG, Robbins KC, Aaronson SA, Antoniades HN (1983) Simian sarcoma virus onc gene, v-sis, is derived from the gene (or genes) encoding a platelet-derived growth factor. Science 221:275-277

Downward J, Yarden Y, Mayes E, Scrace G, Totty N, Stockwell A, Ullrich J, Schlessinger, Waterfield MD (1984) Close similarity of epidermal growth factor receptor and v-erb-B oncogene protein sequences. Nature (London) 307:521-527

Ebina Y, Ellis L, Jarnagin K, Edery M, Graf L, Clausen E, Ou J, Masiazz F, Kan YW, Goldfine ID, Roth RA, Rutter WJ (1985) The human insulin receptor cDNA: the structural basis for hormone-activated transmembrane signalling. Cell 40:747-758

Edwards DK, Parfett CLJ, Denhardt DT (1985) Transcriptional regulation of two serum-induced RNAs in mouse fibroblasts: equivalence of one species to B2 repetitive elements. Mol Cell Biol 5:3280-3288

Eliyahu D, Raz A, Gruss P, Givol D, Oren M (1984) Participation of p53 cellular tumor antigen in transformation of normal embryonic cells. Nature (London) 312:646-649

Fan K (1982) S-100 protein synthesis in cultured glioma cells i G1-phase of cell cycle dependent. Brain Res 273:498-503

Farrar WL, Anderson WB (1985) Interleukin-2-stimulates association of protein kinase C with plasma membrane. Nature (London) 315:233-235

Feramisco JR, Gross M, Kamata T, Rosenberg M, Sweet RW (1984) Microinjection of the oncogene form of the human H-ras (T-24) protein results in rapid proliferation of quiescent cells. Cell 38:109-117

Gerke V, Weber K (1985) The regulatory chain in the p36-kd substrate complex of viral tyrosine-specific protein kinases is related in sequence to the S-100 protein of glial cells. EMBO J 4:2917-2920

Glenney JR Jr, Tack BF (1985) Amino-terminal sequence of p36 and associated p10. Identification of the site of tyrosine phosphorylation and homology with S-100. Proc Natl Acad Sci USA 82:7884-7888

Greenberg ME, Ziff EB (1984) Stimulation of 3T3 cells induces transcription of c-fos proto-oncogenes. Nature (London) 311:433-438

Hann SR, Thompson CB, Eisenmann RN (1985) c-myc oncogene protein synthesis is independent of the cell cycle in human and avian cells. Nature (London) 314:366-369

Hazelton B, Mitchell B, Tupper J (1979) Calcium, Magnesium, and growth control in the WI-38 human fibroblasts cell. J Cell Biol 83:487-498
Hirschhorn RR, Aller P, Yuan Z-A, Gibson CW, Baserga R (1984) Cell cycle specific cDNAs from mammalian cells temperature-sensitive for growth. Proc Natl Acad Sci USA 81:6004-6008
Hyland JK, Rogers CR, Scolnick EM, Stein RB, Ellis R, Baserga R (1985) Microinjected ras family oncogenes stimulate DNA synthesis in quiescent mammalian Cells. Virology 141:333-336
Isobe T, Okuyama T (1981) The amino-acid sequences of the subunit in bovine brain S-100a protein. Eur J Biochem 16:79-86
Isobe T, Ishioka N, Okuyama T (1981) Structural relation of two S-100 protein in bovine brain; subunit composition of S-100a protein. Eur J Biochem 115:469-474
Kahn HJ, Marks A, Thom H, Baumal R (1983) Role of antibody to S-100 protein in diagnostic pathology. Am J Clin Pathol 79:341-347
Kaczmarek L, Hyland JK, Watt R, Rosenberg M, Baserga R (1985a) Microinjected c-myc as a competence factor. Science 228:1313-1315
Kaczmarek L, Calabretta B, Baserga R (1985b) Expression of cell-cycle-dependent genes in phytohemagglutinin-stimulated human lymphocytes. Proc Natl Acad Sci USA 82: 5375-5379
Kaczmarek L, Oren M, Baserga R (1986) Co-operation between the p53 protein tumor antigen and platelet-poor plasma in the induction of cellular DNA synthesis. Exp Cell Res 162:268-272
Kaibuchi K, Takay Y, Nishizuka Y (1985) Protein kinase C and calcium ion in mitogenic response of macrophage-depleted human peripheral lymphocytes. J Biol Chem 260: 1366-1369
Kelly K, Cochran BH, Stiles CD, Leder P (1983) Cell specific regulation of the c-myc gene by lymphocyte mitogens and platelet derived growth factor. Cell 35:603-610
Kruijer W, Cooper JA, Hunter T, Verma IM (1984) Platelet-derived growth factor induces rapid but transient expression of the c-fos gene and protein. Nature (London) 312:711-716
Lane DP, Crawford LV (1979) T antigen is bound to a host protein in SV40 transformed cells. Nature (London) 278:261-263
Lau LF, Nathans D (1985) Identification of a set of genes expressed during the G_0/G_1 transition of cultured mouse cells. EMBO J 4:3145-3151
Lieberman I, Abrams R, Ove P (1963) Changes in the metabolism of ribonucleic acid preceding the synthesis of deoxyribonucleic acid in mammalian cells cultured from the animal. J Biol Chem 238:2141-2149
Linzer DIH, Levine AJ (1979) Characterization of a 54K dalton cellular SV40 tumor antigen present in SV40 transformed cells and uninfected embryonal carcinoma cells. Cell 17:43-52
Linzer DIH, Nathans D (1983) Growth regulated changes in specific mRNAs of cultured mouse cells. Proc Natl Acad Sci USA 80:4271-4275
Linzer DIH, Nathans D (1984) Nucleotide sequence of a growth-related mRNA encoding member of the prolactin-growth hormone family. Proc Natl Acad Sci USA 81:4255-4259
Linzer DIH, Maltzman W, Levine AJ (1979) The SV40 A gene product is required for the production of a 54,000 MW cellular tumor antigen. Virology 98:308-318
Liu HT, Baserga R, Mercer WE (1985) Adenovirus type 2 activates cell cycle-dependent genes that are a subset of those activated by serum. Mol Cell Biol 5:2936-2942
Lorincz AT, Reed SI (1984) Primary structure between the product of yeast cell division control gene CDC 28 and vertebrate oncogenes. Nature (London) 307:183-185
Matrisian LM, Glaichenhaus N, Gesnel MC, Breathnach R (1985a) Epidermal growth factor and oncogenes induce transcription of the same cellular mRNA in rat fibroblasts. EMBO J 4:1435-1440
Matrisian LM, Rautmann G, Magun BE, Breathnach R (1985b) Epidermal growth factor or serum stimulation of rat fibroblasts induces an elevation in mRNA levels for lactate dehydrogenase and other glycolytic enzymes. Nucleic Acids Res 13:711-726
Mercer WE, Baserga R (1985) Expression of p53 protein during the cell cycle of human peripheral blood lymphocytes. Exp Cell Res 160:31-46
Mercer WE, Nelson D, DeLeo AB, Old LJ, Baserga R (1982) Microinjection of monoclonal antibody to protein p53 inhibits serum-induced DNA synthesis in 3T3 cells. Proc Natl Acad Sci USA 79:6309-6312

Milner J, Milner S (1981) SV40 53K antigen: a possible role for 53K in normal cells. Virology 112:785-788
Muller R, Bravo R, Burckhardt J, Curran T (1984) Induction of c-fos gene and protein by growth factors precedes activation of c-myc. Nature (London) 312:716-720
Moore BW (1965) A soluble protein characteristic of the nervous system. Biochem Biophys Res Commun 19:739-744
Nishizuka Y (1984) The role of protein kinase C in cell surface signal transduction and tumour promotion. Nature (London) 308:693-698
Okayama H, Berg P (1983) A cDNA cloning vector that permits expression of cDNA inserts in mammalian cells. Mol Cell Biol 3:280-289
Parfett CLJ, Hamilton RT, Howell BW, Edwards DK, Nilsen-Hamilton M, Denhardt DT (1985) Characterization of a cDNA cloning encoding murine mitogen-regulated protein: regulation of mRNA levels in mortal and immortal cell lines. Mol Cell Biol 5:3289-3292
Powers S, Kataoka T, Fasano O, Goldfarb M, Strathern J, Broach J, Wigler M (1984) Genes in S. Cerevisiae encoding proteins with domains homologous to the mammalian ras proteins. Cell 36:607-612
Raben DM, Cunningham DD (1985) Effects of EGF and thrombin on inositol-containing phospholipids of cultured fibroblasts: stimulation of phosphotidylinositol synthesis by thrombin but not EGF. J Cell Physiol 125:582-590
Reich NC, Levine AJ (1984) Growth regulation of a cellular tumor antigen p53 in nontransformed cells. Nature (London) 308:199-201
Rittling SR, Gibson CW, Ferrari S, Baserga R (1985) The effect of cycloheximide on the expression of cell cycle dependent genes. Biochem Biophys Res Commun 132: 327-335
Rossini M, Baserga S, Huang CH, Ingles CJ, Baserga R (1980) Changes in RNA polymerase II in a cell cycle specific temperature sensitive mutant of hamster cells. J Cell Physiol 103:97-103
Ruscetti SK, Scolnick EM (1983) Expression of a transformation related protein (p53) in the malignant stage of Friend virus-induced diseases. Virol 46:1022-1026
Sawyer ST, Cohen S (1981) Enhancement of calcium uptake and phosphotidylinisitol turnover by epidermal growth factor in A431 cells. Biochemistry 20:6280-6286
Scher CD, Shepard RC, Antoniades HN, Stiles CD (1979) Platelet-derived growth factor and the regulation of the mammalian fibroblasts cell cycle. Biochem Biophys Acta 560:217-241
Sherr CJ, Rettenmier CW, Sacca R, Rousell MF, Look AT, Stanley ER (1985) The c-fms protooncogenes product is related to the receptor for the mononuclear phagocyte growth factor, CSF-1. Cell 41:665-675
Stacey DW, Kung HF (1984) Transformation of NIH 3T3 cells by microinjection of HR-ras p21 protein. Nature (London) 310:508-511
Stefansson K, Wollman R, Jerkovic M (1982) S-100 protein in soft-tissue tumors derived from Schwann cells and melanocytes. Am J Pathol 106:261-268
Sugimoto Y, Whitman M, Cantley LC, Erikson RL (1984) Evidence that the Rous sarcoma virus transforming gene product phosphorylates phosphotidylinositol and diacylglycerol. Proc Natl Acad Sci USA 82:2117-2121
Thompson CB, Challoner PB, Neiman PE, Groudine M (1985) Levels of c-myc oncogene mRNA are invariant throughout the cell cycle. Nature (London) 314:363-366
Torelli G, Selleri L, Donelli A, Ferrari S, Emilia G, Venturelli D, Morelli L, Torelli U (1985) Activation of c-myb expression by phytohemagglutinin stimulation in normal human T lymphocytes. Mol Cell Biol 5:2874-2877
Van Eldik LJ, Zendegui JG, Marshak DR, Watterson DM (1982) Calcium-binding proteins and the molecular basis of calcium action. Int Rev Cytol 77:1-61
Whitman M, Kaplan DR, Schaffhausen B, Cantley L, Roberts TM (1985) Association of phosphotidylinositol and kinase activity with polyoma middle-T component for transformation. Nature (London) 315:239-242

Cell Cycle Control Genes of the Fission Yeast *Schizosaccharomyces pombe*

V. Simanis, P. Russell, and P. Nurse[1]

Yeasts provide an excellent model system for the study of the basic mechanisms of eukaryotic cell cycle control. Since they are simple eukaryotes, cell cycle regulation is less likely to be confused with the additional controls involved in growth and differentiation which operate in muticellular organisms.

The cell cycle comprises all of the events that occur between the birth of a cell and its subsequent division into two daughters. Taken literally, this includes not only those processes required for cellular reproduction, but also those involved in cellular growth. The ease of genetic analysis in yeast has made possible the isolation of many conditionally lethal mutants which under nonpermissive conditions are unable to complete the cell cycle (see for example, Hartwell et al. 1973, Nurse et al. 1976, Nasmyth and Nurse 1981). Of these mutants, those which continue growth and macromolecular synthesis at the restrictive temperature may be considered to be specifically defective in functions required for cell reproduction rather than cell growth. By this criterion about 40 cell division cycle *(cdc)* genes have been identified in the fission yeast *Schizosaccharomyces pombe*. Analysis of mutants in these *cdc* genes have identified two control points in the cell cycle.

Cell Cycle Control Points in Fission Yeast

The cell cycle of *S. pombe* is typically eukaryotic with discrete G1, S, G2, and M phases. Two major control points have been identified in the fission yeast cell cycle. First, a control located in G1 termed "Start". Once Start has been traversed, cells become committed to the mitotic cell cycle in the sense that they are then unable to undergo the alternative development fate of conjugation and sporulation. Temperature-sensitive lethal mutants in two genes, *cdc2*$^+$ and *cdc10*$^+$ block at a point in the cell cycle during G1 where cells are still able to conjugate if challenged to do so (Nurse and Bissett 1981). Traverse of Start in fission yeast requires cells to attain a critical size and in many growth conditions appears to be a major rate-limiting step of the cell cycle (Fantes 1977, Nurse and Thuriaux 1977, Nasmyth 1979, Nasmyth et al. 1979).

A second control point acts in the G2 period and determines the timing of the initiation of mitosis. This mitotic control was revealed by the isolation of *wee* mutants which undergo mitosis and cell division at about half the size of wild-type cells (Nurse 1975, Thuriaux et al. 1978). These "wee" mutants are advanced through G2 into mitosis and thus define functions that are rate-limiting for the traverse of G2 and are involved in determining the timing of mitosis. The wee mutants

[1]Cell Cycle Control Laboratory, Imperial Cancer Research Fund, PO Box 123, Lincoln's Inn Fields, London, WC2A 3PX, England

37. Colloquium Mosbach 1986
Cell Cycle and Oncogenes

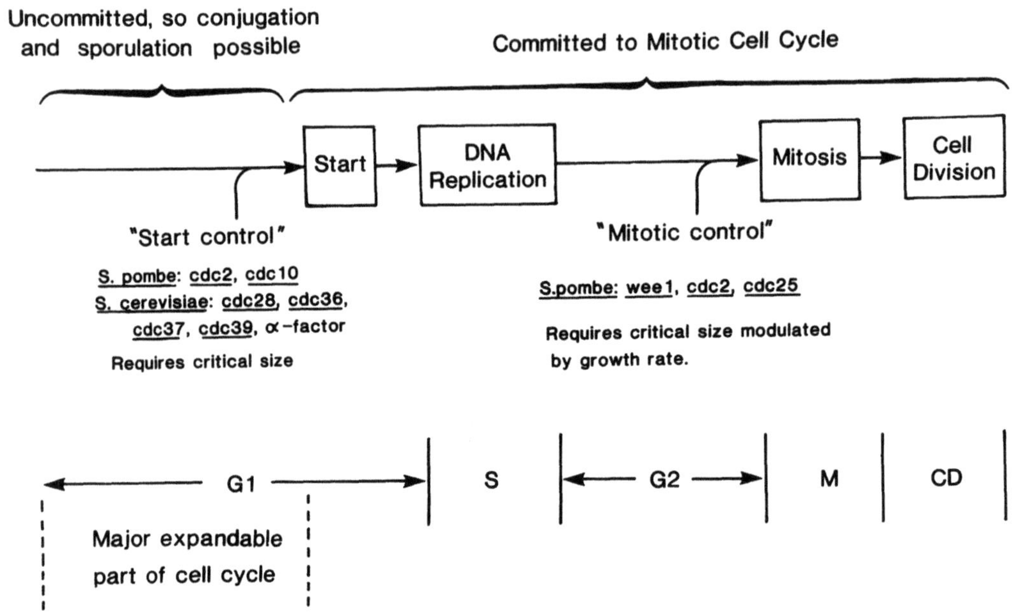

Fig. 1. Summary of cell cycle controls. The start and mitotic controls are marked. Start divides the cell cycle into an uncommitted phase and a committed phase. The major genes involved are given under the respective controls

were shown to map to two genes: *wee1*$^+$ and *cdc2*$^+$ (Nurse and Thuriaux 1980). Genetic analysis indicated that the *wee1*$^+$ gene encodes a dosage dependent inhibitor of mitosis while the *cdc2*$^+$ gene encodes an inducer of mitosis. Mutations in the *cdc2* gene can thus be of two types. One type is recessive, temperature-sensitive lethal mutations causing cells to arrest either in G1 or in late G2 and the other type named *cdc2w* are dominant *wee* mutations causing cells to initiate mitosis prematurely.

There is evidence suggesting that a third gene, *cdc25*$^+$, is also involved in the mitotic control in *S. pombe*. Temperature sensitive lethal *cdc25*$^{t.s}$ mutants become blocked in late G2 upon shift to the restrictive temperature and are unable to initiate mitosis (Thuriaux et al. 1980). This defect can be suppressed both by *wee1*$^-$ and *cdc2w* mutations (Fantes 1979, 1981) suggesting that the *wee1*$^+$, *cdc2*$^+$ and *cdc25*$^+$ gene products might interact to determine the timing of mitosis. The genes acting at the two control points in the mitotic cell cycles are summarized in Fig. 1. It is striking that one gene, *cdc2*$^+$, is uniquely involved at both major control points, acting first in G1 at the point of commitment to the mitotic cell cycle, and then again in G2, being involved in the mechanism which determines the timing of mitosis.

In this paper we will review the cloning and analysis of the *cdc2*$^+$, *cdc10*$^+$ and *cdc25*$^+$ genes. Characterization of the *cdc25*$^+$ gene has demonstrated that it acts as a dosage-dependent inducer in the mitotic control, functioning to counteract the inhibition of mitosis caused by the *wee1*$^+$ gene product. Isolation and characterization of the *cdc2*$^+$ gene has shown that it encodes a protein kinase whose activity may be regulated by phosphorylation. This has led to the hypothesis that phosphorylation of a specific set of proteins may control the G1-S and G2-M transitions, and that modulation of the cdc2 protein kinase activity may be a part of the mechanism by which the cell regulates exit from the mitotic cell cycle and entry into stationary phase.

Cloning and Characterization of the Cell Cycle Genes *cdc2*, *cdc10* and *cdc25*

Plasmids containing the *cdc2*$^+$ (Beach et al. 1982), *cdc10* (Aves et al. 1985) and *cdc25*$^+$ (Russel and Nurse 1986) genes have been isolated from gene banks by virtue of their ability to permit the growth of the *cdc*$^{t.s.}$ mutant strain at the restrictive temperature. The cloned DNA's were in all cases shown to contain these genes by integrating the plasmids via homologous recombination into the chromosome, and then showing that plasmid-linked markers were now closely linked to the chromosomal sites of the genes. Extragenic suppressors of *cdc2*$^{t.s}$ and *cdc25*$^{t.s}$ were also isolated and are currently being characterized.

The *cdc2*$^+$ gene was initially identified on a 4.7 kb *Hind*III fragment and subsequently localized by a combination of subcloning and *Bal*31 resection to a 2 kb fragment which specifies a 1.6 kb polyadenylated transcript (Durkacz et al. 1985, 1986). Given the central role of the *cdc2*$^+$ gene function in control of the cell cycle, we sought to investigate whether changes in *cdc2*$^+$ transcript were responsible for regulating entry and passage through the mitotic cell cycle. A culture of synchronous cells was prepared by selecting small cells at the beginning of the cell cycle using an elutriator rotor, and RNA was prepared at various times. The level of *cdc2* transcript was analyzed by Northern blotting. Two mitotic cell cycles were sampled and no significant change in the steady state level of the *cdc2*$^+$ transcript were observed (Fig. 2). No change in the level of *cdc2*$^+$ transcript was found as cells underwent the transition from cell proliferation to stationary phase (Durkacz et al. 1986). A similar result has been obtained for the *cdc10*$^+$ gene which encodes a 2.7 kb polyadenylated transcript (Aves et al. 1985). The steady state level of this transcript is unaltered during synchronous cultures or during the shift between rapid growth and stationary phase. Thus, changes in the steady state levels of the *cdc2*$^+$ and *cdc10*$^+$ gene transcripts do not play any role in regulating passage through or exit from the mitotic cell cycle. This is consistent with observations showing that about a 10 - 20-fold increase in *cdc2*$^+$ and *cdc10*$^+$ mRNA levels has no effect on the cell cycle.

The *cdc2*$^+$ *cdc10*$^+$, and *cdc25*$^+$ genes have been sequenced and have the potential to encode proteins of MW. 34,000, 85,000, and 70,000 respectively (Hindley and Phear 1984, Aves et al. 1985, Russell and Nurse 1986). Computer searches failed to identify any proteins with significant similarities to the predicted *cdc10*$^+$ or *cdc25*$^+$ proteins. However, a number of homologies were identified with the predicted *cdc2*$^+$ protein. At the amino acid level, about 20% homology is seen with the catalytic subunit of cAMP-dependent protein kinase and a number of putative protein kinases from the *src* family of oncogenes. A much higher level of homology, about 62%, was found with the cell division cycle gene *CDC28* of the budding yeast *Saccharomyces cerevisiae* (Hindley and Phear 1984, Nurse 1985).

In order to maximize the homology between the *cdc2*$^+$ and *CDC28* gene reading frames, it was necessary to propose the existence of four intervening sequences in the *cdc2*$^+$ gene. The *CDC28* gene has no intervening sequence (Lorincz and Reed 1984). The presence of these four introns was subsequently confirmed by S1 mapping (Durkacz et al. 1986). The *CDC28* gene, like the *cdc2*$^+$ gene, has two points of action during the

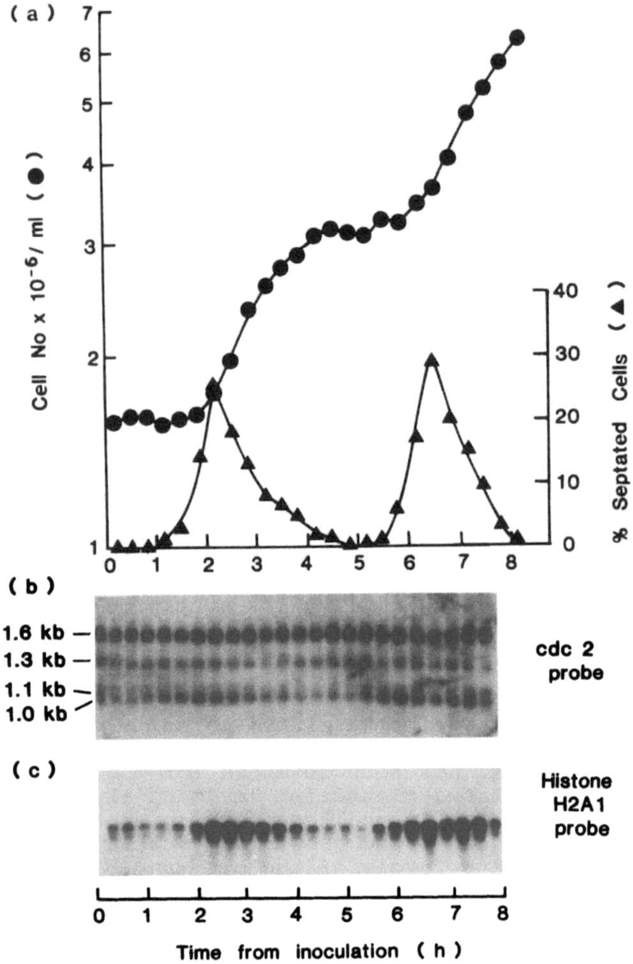

Fig. 2. *cdc2* transcript level during the cell cycle: Small cells were selected from exponentially growing culture using an elutriator rotor, (a) cell number (o) and percentage of septated cells (..) after re-inoculation; (b) 6 µg samples of a total RNA extracted at 20-min intervals after re-inoculation, run on an agarose gel, blotted, probed with the *cdc2* gene and autoradiographed; (c) the same blot stripped and reprobed with histone H2A1 gene (a gift from M. Yanagida). Strain used was *S. pombe* $972h^-$

mitotic cell cycle, being required in G1 at Start, again later in the cell cycle to initiate mitosis (Piggott et al. 1982). The *cdc2*$^+$ and *CDC28* genes are closely related with respect to their molecular function. A *CDC28* gene introduced by transformation into a *S. pombe cdc2*$^{t.s}$ mutant will rescue the mutant function and permit the cells to complete the cell cycle at the restrictive temperature, demonstrating the near functional equivalence of these two gene products (Beach et al. 1982).

Comparison of the regions concerved between *cdc2*$^+$ and *CDC28* allows the significance of the approximately 20% homology observed between cdc2 and other protein kinases to be assessed. This analysis shows only 8% identity between the sequences of *cdc2*$^+$, *CDC28*, v-*mos* and the catalytic subunit of cAMP-dependent protein kinase, at first sight, not very significant. However, most of this similarity may be accounted for the presence of two functionally important regions within the proteins; an ATP binding site of the sort found in protein kinases and a phosphorylation site (Nurse 1985).

Analysis of the *cdc2⁺* Gene Product

The similarities between *cdc2*$^+$ and certain protein kinases strongly suggest that the *cdc2*$^+$ gene product is a protein kinase and probably also a phosphoprotein. To confirm this and to investigate *cdc2*$^+$ function we have prepared *cdc2*$^+$ specific antibodies. After removal of the four intervening sequences, the *cdc2*$^+$ gene has a single long open reading frame 297 amino acids in length (Hindley and Phear 1984). Three peptides were synthesized from the predicted amino acid sequence of the protein, and after coupling to a carrier protein they were used to immunize rabbits. The resulting heteroantisera detect a *S. pombe* protein of 34 kDa (p34). The level of this protein is greatly increased by fusion of the *cdc2*$^+$ to a strong promoter, showing that it is *cdc2*$^+$ protein specific (Simanis and Nurse 1986).

Immunoprecipitation from cell extracts labeled with ^{32}P has demonstrated that the *cdc2*$^+$ protein is phosphorylated in vitro. Immune complex kinase assays were then performed, and established that the immunoprecipitates which contain *cdc2*$^+$ protein possess protein kinase activity, which is thermolabile when prepared from *cdc2*$^{t.s}$ mutant strains. This strongly suggests that the *cdc2*$^+$ protein is indeed a protein kinase. The possibility that the *cdc2*$^+$ protein is tightly associated with a protein kinase rather than possessing intrinsic protein kinase activity, however, cannot be formally eliminated with the available data.

The *cdc2*$^+$ protein kinase activity was not significantly stimulated by addition of Zn^{2+} to the reaction. This observation is interesting in view of the report by Reed et al. (1985) that the *CDC28* protein kinase activity is activated by Zn^{2+}. The reason for this discrepancy is at present unclear, however, it is not noteworthy that the two putative zinc deletion sites identified in the *CDC28* protein by Reed et al. (1985) are not conserved in the *cdc2*$^+$ protein (Hindley and Phear 1984). This may potentially reflect a difference in the way the two proteins are regulated.

We have monitored the steady state protein levels and gross phosphorylation state of the *cdc2*$^+$ protein in synchronous cultures to investigate whether changes in either of these parameters could play any role in regulating passage through the cell cycle. Synchronous cultures were generated using an elutriator rotor and protein extracts prepared at various times. No significant changes were found in the steady state *cdc2*$^+$ protein level or in its degree of phosphorylation through two mitotic cell cycles (Simanis and Nurse 1986).

To establish whether any changes in the level or phosphorylation state of the *cdc2*$^+$ proteins occurred during exit from the mitotic cell cycle, these parameters were monitored and cells in exponential growth were starved of nitrogen causing them to complete the mitotic cell cycle in progress and accumulate in G1 before Start. As shown in Fig. 3B, no significant changes in protein level were observed as the cells arrested in G1. However, it was found that within 3 h after the shift to nitrogen-free medium, when most cells had accumulated in G1, the *cdc2*$^+$ protein had become dephosphorylated (Fig. 3D). This was paralleled by a decrease of in vitro *cdc2*$^+$ protein kinase activity which had fallen to less than 10% of its initial value by 3 h (Fig. 3C), suggesting that dephosphorylation of the *cdc2*$^+$ protein may lead to a decrease in its protein kinase activity. When starved cells are shifted back into complete medium the protein kinase activity is restored after a 3-h lag period, about the time when the cells would be expected to be traversing Start. Cell division follows 2 - 3 h later, roughly equivalent to a period of one cell cycle (Simanis and Nurse 1986).

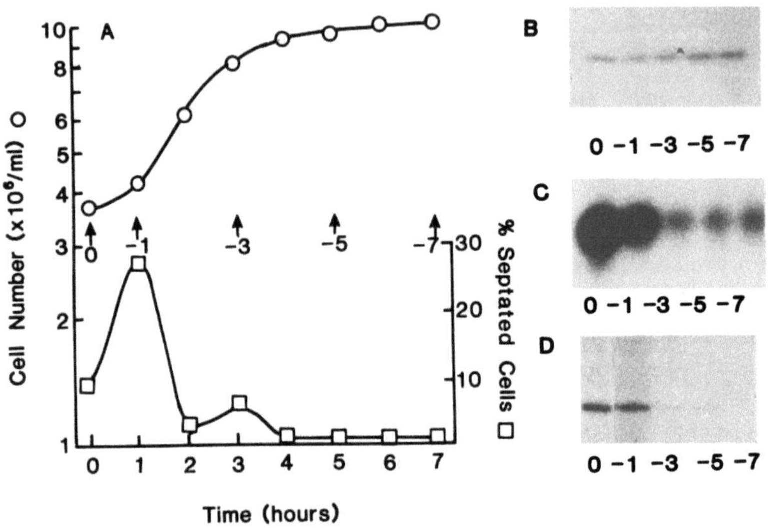

Fig. 3. p34 protein on entry into stationary phase: Exponentially growing cells from an asynchronous culture were shifted into medium lacking nitrogen and protein extracts prepared at various times on the cells entered stationary phase. (A) Cell number and percentage of septated cells after shift to nitrogen-free medium; (B) Protein samples were prepared at the indicated times and analyzed by Western blotting to determine the steady state level of p34; (C) Immunoprecipitates were prepared at the indicated times and assayed for protein kinase activity as described in experimental procedures; (D) Protein extracts were prepared from ^{32}P labeled cultures at the indicated times and p34 immunoprecipitated using serum III. The immune complexes were analyzed on a 10% polyacrylamide gel, which was then dried and autoradiographed for 12 days at -70°C with an intensification screen

Taken together, these findings are consistent with the idea that while the *cdc2⁺* protein remains phosphorylated, the protein kinase activity is maintained and causes cells to remain committed to the mitotic cell cycle. Upon nutrient deprivation the *cdc2⁺* protein becomes dephosphorylated and loses its protein kinase activity, and as a consequence cells are unable to initiate the next cell cycle and accumulate in G1. It is therefore possible that modulation of the *cdc2⁺* protein kinase activity by changes in the phosphorylation state of the protein may be part of the mechanism by which the cell controls the transition from rapid growth to stationary phase. However, our current data do not permit us to exclude other possibilities, for instance that some modification of the *cdc2⁺* protein other than dephosphorylation occurs to effect the decrease in protein kinase activity upon nutrient starvation.

While some insight has been gained into the role of the *cdc2⁺* protein during the transition from rapid growth to stationary phase, no real clues have been obtained concerning the role of the *cdc2⁺* protein in determining the timing of mitosis. No significant changes in the steady state level of the protein or its overall phosphorylation state were observed in synchronous cultures during the G2 period leading up to mitosis. This suggests that the timing of mitosis may be affected by more subtle means, such as changes in availability of substrates for the *cdc2⁺* protein kinase.

The *cdc25*$^{+}$ Gene Encodes a Mitotic Inducer

In order to gain further insight into the mitotic control, we have embarked on a detailed analysis on a second gene involved in determining the timing of mitosis, *cdc25*$^{+}$. The gene has been cloned by rescue of mutant function (Russell and Nurse 1986). DNA sequencing has shown that it could encode a protein of about 70 kDa, although the predicted primary structure has not provided any clues as to its biochemical activity. Since *cdc25*$^{t.s}$ mutations are suppressed by mutations which inactivate the dosage-dependent mitotic inhibitor *wee1*$^{+}$, we investigated whether the *cdc25*$^{+}$ gene might encode a mitotic inducer. We increased the level of *cdc25*$^{+}$ gene expression by integrating multiple copies of the gene into the chromosome at the *cdc25*$^{+}$ locus or by fusing the *cdc25* gene to a strong promoter. These experiments showed that as the level of *cdc25*$^{+}$ gene expression increased, the cell size at division decreased. Increased expression of *cdc25*$^{+}$ did not alter the percentage of binucleate cells or the generation time of the cells. This means that the reduction of the cell size at division resulted from a shortening of the G2 period and a premature initiation of mitosis compared with wild-type cells, suggesting that the *cdc25*$^{+}$ gene product encodes a mitotic induction factor, the execution of whose function is rate-limiting in wild-type cells (Russell and Nurse 1986).

We also investigated the relationship of the *cdc25*$^{+}$ gene with another element important in the mitotic control, the *wee1*$^{+}$ gene. The *wee1*$^{+}$ gene product has been shown by classical genetic techniques to encode an inhibitor of mitosis (Nurse and Thuriaux 1980). Since *cdc25*$^{+}$ gene encodes a mitotic inducer, it is possible that *cdc25*$^{+}$ and *wee1*$^{+}$ could exercise their control by interaction with each other (Fantes 1979). Thus the *wee1*$^{+}$ gene product could block mitosis by inhibiting an essential mitotic induction function of the *cdc25*$^{+}$ gene product. To address this question we have deleted the chromosomal copy of the *cdc25*$^{+}$ gene (Russell and Nurse 1986). In cells that are otherwise wild-type, this deletion is lethal, but in *wee1*$^{-}$ cells the deletion is not lethal. This demonstrates that *wee1* is able to inhibit mitosis by a mechanism which does not involve direct inhibition of *cdc25*$^{+}$ function, and also indicates that the *cdc25*$^{+}$ gene product is only essential for mitotic induction if the mitotic inhibitor *wee1*$^{+}$ is active. These data are consistent with a model which proposes that the *cdc25*$^{+}$ gene product acts to induce mitosis by blocking the inhibitory effect of the *wee1*$^{+}$ gene products. A prediction from such a model is that if the *wee1*$^{+}$ function is inactivated, overexpression of *cdc25*$^{+}$ function should have no additional effect on mitosis to that observed in a *wee1*$^{-}$ mutant alone. We therefore overexpressed the *cdc25*$^{+}$ gene by fusing it to a strong promoter in a strain carrying a temperature sensitive mutation of the *wee1*$^{+}$ gene. At the permissive temperature, when the *wee1*$^{t.s}$ gene product is active, the cells are viable and phenotypically *wee*. However, upon shifting to the restrictive temperature cell division ceased and a rapid loss of cell viability occurred. A number of nuclear and mitotic abnormalities of the following types were observed (Fig. 4): (1) both nuclei partitioning to the same half of a dividing cell, thereby generating binucleate and anucleate cells, (2) cell division without mitosis, generating anucleate cells, (3) bisection of the mitotic spindle by the cell septum, (4) unequal nuclear fission and (5) asymmetrical location of cell plates within the cell (Russell and Nurse 1986). Many cells had cell plates at a cell length of around 5 µm, significantly smaller than the size of *wee* mutants at division, about 8 µm (Nurse 1975). A possible explanation for these phenotypes, which we term mitotic catastrophes, is that mitosis is being initated too prematurely to produce viable cells.

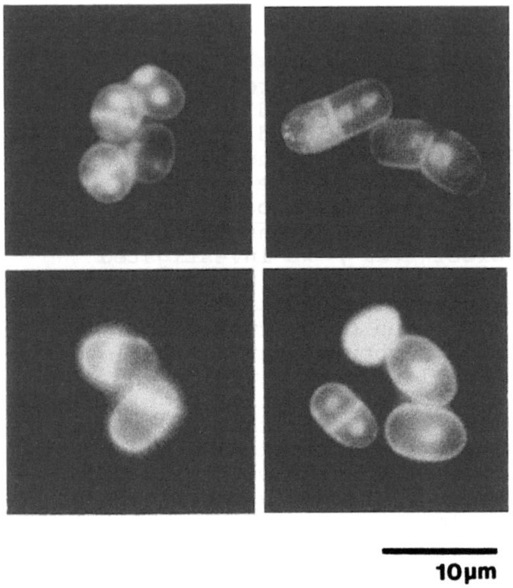

Fig. 4. Abnormal morphologies in *wee1-50* psX-179 integrant strain caused by temperature shift-up to 35°C: A *wee1-50* psX-179 integrant strain was grown to early log phase in minimal medium at 25°C and then shifted to 35°C. These photographs were taken between 2 - 4 h after the temperature shift-up. Nuclear abnormalities are apparent in the three DAPI stained photographs, while the assymetric cell plates are particularly apparent in the Calcofluor photograph *(lower left)*. The smallest cells with cell plates are about 5.0 µm long

These findings demonstrate that the effects of increasing expression of *cdc25⁺* and inactivation of *wee1⁺* function are additive, and thus argue against models which propose that the *cdc25⁺* gene product acts to induce mitosis by direct repression of *wee1⁺* function. This suggests that the *cdc25⁺* and *wee1⁺* gene products exercise their control over the initiation of mitosis independently of each other.

The data presented here lead to a concept of mitotic control by *wee1⁺* and *cdc25⁺* in which the two gene products operate in an antagonistic fashion to regulate the initiation of mitosis. This idea is supported by the observation that inactivation of *cdc25⁺* function is not lethal to the cell if *wee1⁺* is also inactivated, and overexpression of *cdc25⁺* is not lethal provided *wee1* remains active. It therefore seems plausible that controlling the ratio of *wee1* and *cdc25⁺* activity could regulate the initial events of mitosis (Russell and Nurse 1986).

Conclusions

We have cloned and analyzed three genes involved in cell cycle control. *cdc10⁺* acts at the G1/S transition when cells become committed to the mitotic cell cycle, *cdc25⁺* acts at the G2/M transition, being involved in the mechanism which determines the timing of mitosis, and *cdc2⁺* uniquely acts at both of these control points.

Transcriptional regulation of the *cdc2⁺* and *cdc10⁺* genes does not appear to play any role in passage through the mitotic cell cycle or in governing exit from it into stationary phase, as the steady state levels of the transcripts do not change significantly either in synchronous cultures of rapidly growing cells, or in cultures undergoing the transition from rapid proliferation to stationary phase. This is also true for a number of other *cdc* genes of *S. pombe* which have been examined (Johnston et al. 1986 unpublished).

We have shown that the *cdc2*$^+$ gene encodes a protein kinase which is itself a phosphoprotein. We have also reviewed evidence consistent with the idea that modulation of the *cdc2*$^+$ protein kinase activity occurs through changes in the phosphorylation state of the protein. This may be a part of the mechanism by which the cell controls the transition from rapid proliferation to a quiescent stationary phase state. We have shown that the *cdc25*$^+$ gene encodes a mitotic inducer, acting independently of and in an antagonistic fashion to the mitotic inhibitor encoded by the *wee1* gene.

In exponential growth, the *cdc2*$^+$ gene function is required twice during the mitotic cell cycle. It is not immediately apparent why a single gene product should be required at both control points, although such a situation could of course of be exploited to ensure that cells cannot undergo mitosis without first having replicated their DNA. Analysis of the steady state level and phosphorylation state of the *cdc2*$^+$ protein has demonstrated that neither of these parameters changes significantly through the mitotic cycle in exponentially growing wild-type cells. Thus the regulation of *cdc2*$^+$ function through the cell cycle must be effected by other means. It seems reasonable to propose that both the traverse of start and the mitotic control will involve phosphorylation of specific sets of proteins resulting in the inhibition of events leading to DNA replication following G1 and to mitosis following G2. The *cdc2*$^+$ protein, which possesses protein kinase activity and acts at both control points could therefore be responsible for phosphorylation of the necessary proteins to permit DNA replication and subsequently mitosis to proceed. Control of the execution of *cdc2*$^+$ function at these points could then be provided by making the appropriate substrates available to the *cdc2*$^+$ protein.

How *cdc25*$^+$ and *wee1*$^+$ act to determine the timing of mitosis remains unclear. It has been suggested that both the *cdc25*$^+$ and *wee1*$^+$ gene products may act antagonistically upon the same element, *cdc2*$^+$ (Nurse and Thuriaux 1980, Fantes 1981). Clearly, it will be of great interest to determine what the substrates and regulators of the *cdc25*$^+$ *wee1*$^+$ and *cdc2*$^+$ gene products are, and current work in our laboratory is directed toward addressing these problems.

In multicellular eukaryotes, protein phosphorylation has been implicated both in the traverse of both the G1/S and G2/M boundaries. Proteins have been described which become phosphorylated only at mitosis (Davis et al. 1983, Vandre et al. 1984), as have changes in the phosphorylation state of nuclear laminas (Ottario and Gerace 1985) and histones (Bradbury et al. 1974) during the cell cycle. There is also suggestive evidence from mammalian cells, *Physarum* and *Xenopus* oocytes that protein phosphorylation may play a role in mitotic control (Gurley et al. 1975, Inglis et al. 1976, Wu and Gerhart 1980, Halleck et al. 1984). There may therefore be similarities in these higher eukaryotic systems to the situation we have described for *S. pombe*.

References

Aves SJ, Durkacz BW, Carr A, Nurse P (1985) Cloning, sequencing and transcriptional control of the *Schizosaccharomyces pombe cdc10* "start" gene. EMBO J 4:457-463

Beach D, Durkacz B, Nurse P (1982) Functionally homologous cell cycle control genes in budding and fission yeast. Nature (London) 300:706-709

Bradbury EM, Inglis RJ, Matthews HR (1974) Control of cell division by very lysine rich histone (F1) phosphorylation. Nature (London) 247:257-261
Davis FM, Tsao TY, Fowler SK, Rao PN (1983) Monoclonal antibodies to mitotic cells. Proc Natl Acad Sci USA 80:2926-2930
Durkacz B, Beach D, Hayles J, Nurse P (1985) The fission yeast cell cycle control gene *cdc2*$^+$: structure of the *cdc2*$^+$ region. Mol Gen Genet 201:543-545
Durkacz B, Carr A, Nurse P (1986) Transcription of the *cdc2*$^+$ cell cycle control gene of the fission yeast *Schizosaccharomyces pombe*. EMBO J 5:369-373
Fantes P (1977) Control of cell size and cycle time in *Schizosaccharomyces pombe*. J Cell Sci 24:51-67
Fantes P (1979) Epistatic gene interactions in the control of division in fission yeast. Nature (London) 279:428-430
Fantes P (1981) Isolation of cell size mutants of a fission yeast by a new selective method: Characterisation of mutants and implications for division control mechanisms. J Bacteriol 146:746-754
Gurley LR, Walters RA, Tobey RA (1975) Sequential phosphorylation of histone subfractions in the Chinese Hamster cell cycle. J Biol Chem 250:3936-3944
Halleck MS, Lumley-Sapanski K, Reed JA, Iyer AO, Mastro AM, Schlegel RA (1984) Characterisation of protein kinase in mitotic and meiotic cell extracts. FEBS Lett 167:193-198
Hartwell LH, Mortimer RK, Culotti J, Culotti M (1973) Genetic control of cell division in yeast: Genetic analysis of cell division mutants. Genetics 74:267-286
Hindley J, Phear G (1984) Sequence of the cell division gene cdc2 from *Schizosaccharomyces pombe:* patterns of splicing and homology of protein kinases. Gene 31: 129-134
Inglis RJ, Langan TA, Matthews HR, Hardie DG, Bradbury EM (1976) Advance of mitosis by histone phosphokinase. Exp Cell Res 97:418-425
Lorincz AT, Reed SI (1984) Primary structure homology between the product of yeast cell division control gene CDC28 and vertebrate oncogenes. Nature (London) 307: 183-185
Nasmyth K (1979) A control acting over the initiation of DNA replication in the yeast *Schizosaccharomyces pombe*. J Cell Sci 36:215-233
Nasmyth K, Nurse P (1981) Cell division cycle mutants altered in DNA replication and mitosis in the fission yeast *Schizosaccharomyces pombe*. Mol Gen Genet 182: 119-124
Nasmyth K, Nurse P, Fraser RSS (1979) The effect of cell mass on the cell cycle timing and duration of S phase in fission yeast. J Cell Sci 39:215-233
Nurse P (1975) Genetic control of cell size at cell division in yeast. Nature (London) 256:547-551
Nurse P (1985) Cell cycle control genes in yeast. Trends Genet 1:51-55
Nurse P, Bissett Y (1981) Gene required in G1 for commitment to the cell cycle and in G2 for control of mitosis in fission yeast. Nature (London) 292:558-560
Nurse P, Thuriaux P (1977) Controls over the timing of DNA replication during the cell cycle of fission yeast. Exp Cell Res 107:365-375
Nurse P, Thuriaux P (1980) Regulatory genes controlling mitosis in the fission yeast *Schizosaccharomyces pombe*. Genetics 96:627-637
Nurse P, Thuriaux P, Nasmyth K (1976) Genetic control of cell division cycle in fission yeast *Schizosaccharomyces pombe*. Mol Gen Genet 146:167-178
Ottaviano Y, Gerace L (1985) Phosphorylation of the nuclear lamins during interphase and mitosis. J Biol Chem 260:624-632
Piggot JA, Rai R, Carter BLA (1982) A bifunctional gene product involved in two phases of the yeast cell cycle. Nature (London) 248:391-394
Reed SI, Hadwiger JA, Lorincz AT (1985) Protein kinase activity associated with the product of the yeast cell division cycle gene CDC28. Proc Natl Acad Sci USA 82: 4055-4059
Russel PR, Nurse P (1986) *cdc25*$^+$ functions as an inducer in the mitotic control of fission yeast. Cell 45:145-153
Simanis V, Nurse P (1986) The cell cycle control gene *cdc2*$^+$ of fission yeast encodes a protein kinase potentially regulated by phosphorylation. Cell 45:261-268

Thuriaux P, Nurse P, Carter B (1978) Mutants altered in the control co-ordinating cell division with cell growth in the fission yeast *Schizosaccharomyces pombe*. Mol Gen Genet 161:215-220

Thuriaux P, Sipizki M, Fantes P (1980) Genetical analysis of a sterile mutant by protoplast fusion in the fission yeast *Schizosaccharomyces pombe*. J Gen Microbiol 116:525-528

Vandre DD, Davis FM, Rao PN, Borisy GG (1984) Phosphoproteins are components of mitotic microtubule organising centres. Proc Natl Acad Sci USA 81:4439-4443

Wu M, Gerhart J (1980) Partial purification and characterization of the maturation-promoting factor from eggs of *Xenopus laevis*. Dev Biol 79:465-477

Hormonal Control of Cell Division in *Saccharomyces cerevisiae*

D. D. Jenness, A. C. Burkholder, and L. H. Hartwell[1]

Conjugation of haploid MATa and MATα cells of *Saccharomyces cerevisiae* occurs only under conditions that are also conducive to active vegetative growth. Thus, cells must decide whether to undergo mitosis or to conjugate with the cell of opposite mating type. The stage of the cell cycle has been defined at which this decision is made. In most vegative cultures of yeast the majority of the cells are budded and beyond the stage of initiation of DNA replication. A minority of cells are unbudded and most of these are prior to the stage of initiation of DNA synthesis. Zygotes are produced between two haploid cells that are both unbudded (1). Indeed, when vegetative cultures of MATa and MATα cells are mixed together, both cell types accumulate in the unbudded phase of the cell cycle (1).

The arrest of the mitotic cycle in the unbudded phase is a consequence of polypeptide hormones produced constitutively by MATa and MATα cells that act specifically to arrest cell division in the cell of opposite mating type (2,3). These hormones have been isolated and sequenced. The α-factor produced by MATα cells contains 13 amino acids (4) and the a-factor produced by MATa cells contains 11 amino acids (5). Both hormones are encoded by two sets of unlinked genes, each set coding for a tandem array of hormone precursors which must be proteolytically processed to active hormones (6,7). The precise stage of cell division arrest by these hormones has been defined by comparing this arrest with that of temperature-sensitive *cdc* mutants using the strategy of reciprocal shifts. Both MATa and MATα cells are arrested by their respective pheromone at the first of three known steps in G1 (8). This is the same stage at which cells arrest when they are nutritionally deprived (reviewed in (9)). Because of the critical role of this step in controlling cell division both with respect to nutritional factors and mating hormones, this step has been termed Start (10). Execution of this step in the normal cell cycle requires genes that are involved in growth regulation: *cdc19* (pyruvate kinase (11)) *cdc35* (adenylate cyclase (12)) *cdc25*, and *cdc33*, as well as the function of some genes that, when deficient, mimic arrest by mating factor: *cdc28* (serine protein kinase (13), *cdc36*, *37*, and *39* (14)).

Completion of the events required for start results in the duplication of the spindle pole body (15,16), an electron dense disk embedded in the continuity of the nuclear membrane which serves as the pole of the mitotic spindle. As a consequence of the arrest of division at this stage in the cell cycle by mating hormones, the two conjugating cells fuse to form a zygote which contains two nuclei each of which has a single spindle pole body. Fusion of the two haploid nuclei to produce a diploid nucleus occurs by fusion between the two single spindle pole bodies to produce a single large spindle pole body characteristic of a diploid cell in G1 (15). Hence the diploid cell begins its mitotic cycle at Start and proceeds through the mitotic cell cycle by first duplicating its spindle pole body.

[1]Department of Genetics SK-50, University of Washington, Seattle, WA 98195, U.S.A.

37. Colloquium Mosbach 1986
Cell Cycle and Oncogenes

The action of the mating hormones upon cell division accounts for the accumulation of cells in G1 in mixed vegetative cultures of MATa and MATα cells and the fact that zygotes are formed between G1 haploid cells. It was not clear from these results whether cells could conjugate at other stages in the cell cycle since the presence of the mating hormones precludes this possibility. We tested whether cells at other stages could conjugate by arresting cells at different stages in the cycle using *cdc* mutations and then challenging them to conjugate from this arrest point. Mutants arrested at all stages other than Start were incapable of mating with cells of opposite of mating type to form a diploid cell (17). Only those arrested at Start by the *cdc28* mutation were capable of conjugating. Thus not only do mating hormones facilitate conjugation at Start but some mechanism operates to prohibit conjugation from other positions in the cell cycle. This fact has been used as a selection scheme for isolating additional start mutants (14).

In addition to the arrest of cell division at Start mating factors induce other responses in their target cell. Alpha factor induces the synthesis of an agglutinin on the surface of MATa cells (18,19,20) and the formation of a projection on the surface of the cell (21), producing a change in cell morphology termed a shmoo. We have investigated these responses of the MATa cell as a function of the dose of α-factor. Agglutinin is induced at a dose of 10^{-10} M α-factor, cell division arrest at a dose of 2.5×10^{-10} M α-factor and projections at a dose of 1.4×10^{-8} M α-factor. The different dose required for agglutination and cell division arrest compared to that required for projection formation is teleologically sensible. At low α-factor concentrations produced when MATα cells are in the same environment as MATa cells, the MATa cells arrest division in G1 and synthesize a surface agglutinin in preparation for conjugation. Upon physical interaction of a MATa cell with a MATα cell, agglutination takes place and now the close proximity of the MATα cell should result in a much higher concentration of α-factor. The high concentration would now induce projection formation, which in essence allows the MATa cell to grow into the MATα and produce a conjugation tube. Although MATa cells will produce projections in an isotropic environment containing α-factor, when a concentration gradient of α-factor exists each cell forms its projection in the direction of the concentration gradient (21). MATα cells produce an enzyme which destroys α-factor (23,24); this activity is controlled by or encoded by BAR 1 gene (25). The function of this enzyme may be to accentuate the concentration gradient of α-factor.

How does the MATa cell produce different responses at different concentrations of the same polypeptide hormone? Are there different receptors with different affinities? Is there one receptor that produces different signals at different occupancies? Is there one receptor that produces one signal and the cell responds differently to different intensity of signal? To approach the answers to these questions we studied the binding of radiolabeled α-factor to MATa cells. Under certain conditions α-factor binds specifically and reversibly to the surface of MATa cells (26,27). Scatchard analysis of the binding data reveals a saturable receptor present at about 8,000 sites per cell with a disassociation constant of about 6×10^{-9} M α-factor. The dose response for binding to the receptor is similar to the dose response for projection formation. The binding data does not rule out the possible presence of other receptors present at lower concentration.

In order to attempt to further resolve the nature of the receptor we have studied the genetic basis of MATa cell responsiveness to α-factor. Four hundred temperature-sensitive mutants resistant to division arrest by α-factor were isolated and studied (28). These mutants defined eight complementation groups. Representative mutants from each complementation group were analyzed for their response to α-factor. Mutations in all eight complementation groups were deficient for cell division arrest, agglutinin induction, and projection formation in response to α-factor. In order to determine which of these genes were involved specifically in the cell's hormone reception and signal transduction system, we also studied the constitutive MATa cell functions in these mutants. That is, those functions expressed specifically by MATa cells that are not inducible by α-factor. These are the production of a-factor and the destruction of α-factor. Mutations in STE 4, 5, 7, 11, and 12 had depressed levels of a-factor production and α-factor destruction (28). Furthermore, mutations in STE 4, 5, 7, 11, and 12 also result in a sterile phenotype in both MATa and MATα cells. Because of their effects on constitutive expression and their involvement in fertility in both mating types, it is unlikely that any of these genes are involved in the α-factor receptor or signal transduction system. Mutations in one gene, STE2, have the expected phenotype for a defect in α-factor receptor. Ste2 mutants produced normal levels of a-factor and α-factor destruction and mutants in STE2 were sterile as MATa cells but fertile as MATα cells.

In agreement with these results, we found that receptors on the surface of temperature-sensitive ste2 mutant cells were thermo-labile compared to wild-type receptors, but not those from mutants in any of the other STE genes (27). When ste2 mutants were grown at the permissive temperature, washed and inactivated for various periods of time at 49°, ligand binding was labile in the mutant cells but not in wild-type cells. These results indicate that the STE2 gene codes for a component of the α-factor receptor.

The STE2 gene was cloned by complementation of the mating defect in temperature-sensitive ste2 mutants (29,30). Complementing activity was obtained on a 4.3 kb restriction fragment; this restriction fragment directed the integration of a plasmid containing it to the STE2 locus, demonstrating that the DNA was homologous with the appropriate locus on the genetic map. Furthermore, integration of this sequence restored mating as well as thermo-resistant ligand binding activity to the cell surface. The sequence of the STE2 locus reveals an open reading frame of 431 amino acids (29,30). The most notable feature of the amino acid sequence is the presence of seven hydrophobic domains, each approximately 40 amino acids long, that constitute the amino terminal two thirds of the coding sequence. The remaining one third at the carboxyl terminus is hydrophilic. These results suggest that the STE2 gene codes for an integral membrane protein.

Having the STE2 gene cloned permits a rigorous test of the role of this gene product in the various α-factor inducible responses of MATa cells. The STE2 coding sequence was replaced by the coding sequence for the LEU2 gene in a plasmid (Burkholder, A.C. and L. Hartwell in preparation). A restriction fragment containing the LEU2 substitution of the STE2 gene was cut out of the plasmid and transformed into yeast in order to transplant the STE2 locus with the LEU2 gene. This manipulation will remove the STE2 coding sequence from the genome. Such transformants were obtained and were verified by Southern blot analysis to have the STE2 gene replaced by the LEU2 gene. These transformants were viable, demonstrating that STE2 gene product is not essential for viability. The transformants were sterile as MATa cells but fertile as MATα cells, demonstrating that the STE2 gene is needed for fertility only in the MATa cell. In MATa cells this deletion eliminates all three α-factor inducible responses: agglutination, cell division arrest, and projection formation. This result demonstrates that the STE2 gene product is needed for all three responses. This result does not prove that a single receptor is involved in all three responses, but it does indicate that if more than one receptor is involved, the STE2 gene product is likely to be a component of all receptors.

In order to further explore the relationship of the different inducible responses, we sought mutants which separate the ability of the cell to exhibit different responses (Burkholder, A.C. and L. Hartwell in preparation). One thousand mutants resistant to division arrest by α-factor were selected. These mutants were then tested to see if any retained the ability to agglutinate. Only approximately 20 mutants retained this capacity. Most of these are alleles of the STE2 gene. Two types were obtained and one representative of each type was studied in some detail. Mutant allele ste2-9 produces agglutinin in the absence of stimulation by α-factor. This constitutive production of agglutinin is due to the mutation at the STE2 locus, since the phenotype is tightly linked in genetic crosses to the STE2 locus. The existence of this mutant demonstrates that the induction of agglutinin is due to a signal that can be produced by the STE2 gene product in the absence of α-factor. Thus the α-factor polypeptide does not form an essential part of the signal-producing domain of the receptor.

Mutants containing the ste2-8 allele remain inducible for agglutinin synthesis by α-factor. The dose response for induction of agglutinin is shifted only slightly from the wild-type curve to a half maximum dose response, about three to four times higher in α-factor concentration. This mutant exhibits little or no response for cell division arrest with increasing concentrations of α-factor, no induction of projections and little or no detectable α-factor binding activity on the surface. Thus, it is possible to obtain mutant alleles of STE2 which retain a nearly normal dose-response for agglutinin induction, but eliminate the responses for cell division arrest, projection formation, and drastically reduce the number of ligand binding sites. The existence of this mutant suggests that the simplest model of a single receptor with a single ligand binding site that produces different responses in direct proportion to the number of occupied receptors may be incorrect.

References

1. Hartwell Leland H (1973) Synchronization of haploid yeast cell cycles, a prelude to conjugation. Exp Cell Res 76:111-117
2. Bucking-Throm E, Duntze W, Hartwell LH, Manney TR (1973) Reversible arrest of haploid yeast cells at the initation of DNA synthesis by a diffusible sex factor. Exp Cell Res 76:99-110
3. Wilkinson LE, Pringle JR (1974) Transient G1 arrest of *S. cerevisiae* cells of mating type α by a factor produced by cells of mating type a. Exp Cell Res 89: 175-187
4. Stotzler D, Kiltz H, Duntze W (1976) Primary structure of α-factor peptides from *Saccharomyces cerevisiae*. Eur J Biochem 69:397-400
5. Betz R, Duntze W (1979) Purification and partial characterization of α-factor, a mating hormone produced by mating-type a cells from *Saccharomyces cerevisiae*. Eur J Biochem 95:469-475
6. Kurjan J, Herskowitz I (1982) Structure of a yeast pheromone gene (M F α): A putative α-factor precursor contains four tandem copies of mature α-factor. Cell 30:933-944
7. Brake AJ, Brenner C, Najarian R, Laybourn P, Merryweather J (1985) Structure of yeast peptide mating pheromone a factor. In: Grethin MJ (ed) Protein transport and secretion. Cold Spring Harbor Labs, Cold Spring Harbor, New York, pp.103-108

8. Hereford Lynna M, Hartwell LH (1974) Sequential gene function in the initiation of Saccharomyces cerevisiae DNA synthesis. J Mol Biol 84:445-461
9. Pringle John R, Hartwell Leland H (1981) The *Saccharomyces cerevisiae* cell cycle. In: Strathern JN, Jones EW, Broach JR (eds) Molecular biology of the yeast *Saccharomyces*, vol I. Cold Spring Harbor Laboratory, Cold Spring Harbor, New York, pp. 97-142
10. Hartwell Leland H, Culotti Joseph, Pringle John R, Reid Brian J (1974) Genetic control of the cell division cycle in yeast. Science 183:46-51
11. Kawsiaki G, Fraenkel DG (1982) Cloning of yeast glycolysis genes by complementation. BBRC 108:1107-1112
12. Boutelet F, Petitjean A, Hilger F (1985) Yeast cdc35 mutants are defective in adenylate cyclase and are allelic with cyrl mutants while CAS1, a new gene, is involved in the regulation of adenylate cyclase. EMBO J 10:2635-1641
13. Reed S, Hadwiger J, Lorincz A (1985) Protein kinase activity associated with the product of the yeast cell division cycle gene CDC28. Proc Natl Acad Sci USA 82:4055-4059
14. Reed S (1980) The selection of *S. cerevisiae* mutants defective in the start event of cell division. Genetics 95:561-577
15. Byers B, Goetsch L (1974) Duplication of spindle plaques and integration of the yeast cell cycle. Cold Spring Harbor Symp Quant Biol 38:123
16. Byers B, Goetsch L (1975) Behavior of spindles and spindle plaques in the cell cycle and conjugation of *Saccharomyces cerevisiae*. J Bacteriol 124:511
17. Reid Brian J, Hartwell Leland H (1977) Regulation of Mating in the Cell Cycle of *Saccharomyces cerevisiae*. J Cell Biol 75:355-365
18. Sakai K, Yanagishima N (1972) Mating reaction in *Saccharomyces cerevisiae* II Hormonal regulation of agglutinability of a type cells. Arch Microbiol 84:191-198
19. Betz R, Duntze W, Manney T (1978) Mating factor mediated sexual agglutination in *Saccharomyces cerevisiae*. FEMS Microbiol Lett 4:107-110
20. Fehrenbacher G, Perry K, Thoner J (1978) Cell-cell recognition in *Saccharomyces cerevisiae:* Regulation of mating - specific adhesion. J Bacteriol 134:893-901
21. Levi JD (1956) Mating reaction in yeast. Nature (London) 177:753-754
22. Moore Susan A (1983) Comparison of dose-response curves for α-factor-induced cell division arrest, agglutination, and projection. J Biol Chem 258:13849-13856
23. Maness PF, Edelman GM (1978) Inactivation and chemical alteration of mating factor α by cells and spheroplasts of yeast. Proc Natl Acad Sci USA 75:1304-1308
24. Ciejek E, Thorner J (1979) Recovery of *S. cerevisiae* a cells from G1 arrest by α-factor pheromone requires endopeptidase action. Cell 18:623-635
25. Manney T (1983) Expression of the BAR1 gene in *Saccharomyces cerevisiae:* induction by the α mating pheromone of an activity associated with a secreted protein. J Bacteriol 155:291-301
26. Thorner J (1980) Intercellular interactions of the yeast *Saccharomyces cerevisiae*. In: Leighton TJ, Loomis WA Jr (eds) Molecular genetics of development: An introduction to recent research on experimental systems. Academic Press, London New York, p 119
27. Jenness DD, Burkholder AC, Hartwell LH (1985) Binding of α-factor pheromone to yeast a cells: Dissociation constant and number of binding sites. Mol Cell Biol 6:318-320
28. Hartwell Leland H (1980) Mutants of *Saccharomyces cerevisiae* unresponsive to cell division control by polypeptide mating hormone. J Cell Biol 85:811-822
29. Burkholder AC, Hartwell LH (1985) The yeast α-factor receptor: Structural properties deduced from the sequence of the *STE2* gene. Nucleic Acids Res 13: 8463-8475
30. Nakayama N, Miyajima A, Arai K (1985) Nucleotide sequences of *STE2* and *STE3*, cell type-specific sterile genes from *Saccharomyces cerevisiae*. EMBO J 4:2643-2648

Molecular Cloning and Physiological Analysis of the Start Gene *cdc25* in Budding Yeast

L. Alberghina, M. Baroni, S. Livian, G. Frascotti, and E. Martegani[1]

Introduction

There is increasing interest for the yeast *Saccharomyces cerevisiae* in studies on the regulatory mechanisms of cell proliferation, since it is particularly well characterized from the biochemical and genetic point of view and furthermore the recombinant DNA techniques have now greatly enhanced the power of classical yeast genetics (Hinnen et al. 1978). *S. cerevisiae* presents a major cell cycle control function in G1, in the regulatory area called Start (Hartwell 1974, Hartwell et al. 1974). At Start an yeast cell integrates many intracellular and environmental signals, and then it is committed to continue proliferation or switched to differentiative pathways such as sporulation, conjugation, or to entry in stationary phase. Growth to a critical cell size appears to be required among other conditions to traverse Start and to be committed to DNA replication and cell division (Nurse 1981, Pringle and Hartwell 1981).

Information on the Start functions have been obtained through the biochemical and genetic analysis of a number of cell-division cycle *(cdc)* mutants that arrest at or near Start (Hartwell et al. 1973, Reed 1980). They have been divided in two classes according to their terminal phenotype (Reed 1980). Class I mutants (like *cdc28*, *cdc36*, *cdc37*, *cdc39*) continue cytoplasmic growth at restrictive temperature, while bud emergence and DNA replication are blocked (Hartwell et al. 1973, Reed 1980). They resemble alpha-factor-treated cells and are able to conjugate at restrictive temperature, but not to sporulate, if diploid. Many Class I Start genes have been cloned and characterized at molecular level (Reed et al. 1982, Lorincz and Reed 1983).

Class II Start mutants (such as *cdc25*, *cdc33*, *cdc35*) arrest at restrictive temperature as round unbudded cells in which a growth lesion is also apparent. They are not able to conjugate at restrictive temperature, but they can sporulate if diploid (Hartwell et al. 1973, Reed 1980). In addition, mutants *cdc25* and *cdc35* behave as sporulation derepressed *(spd)* mutants (Shilo et al. 1978, Dawes and Calvert 1984). Although the biochemical and genetic studies carried out so far do not clearly indicate a biochemical function for the *CDC25* gene product, several observations suggest the working hypothesis that the *CDC25* gene product could be part of a mechanism that, in response to internal and/or external signals, controls the developmental pathways in yeast, and that this product interacts, directly or indirectly, with the cAMP pathway (Martegani et al. 1986, Camonis et al. 1986). In the present paper we report several observations related to the molecular analysis of the *CDC25* gene, as well as to its transcriptional regulation and physiological role.

[1]Dipartimento di Fisiologia e Biochimica Generali-Sezione di Biochimica Comparata, Università di Milano, Milano, Italy

37. Colloquium Mosbach 1986
Cell Cycle and Oncogenes

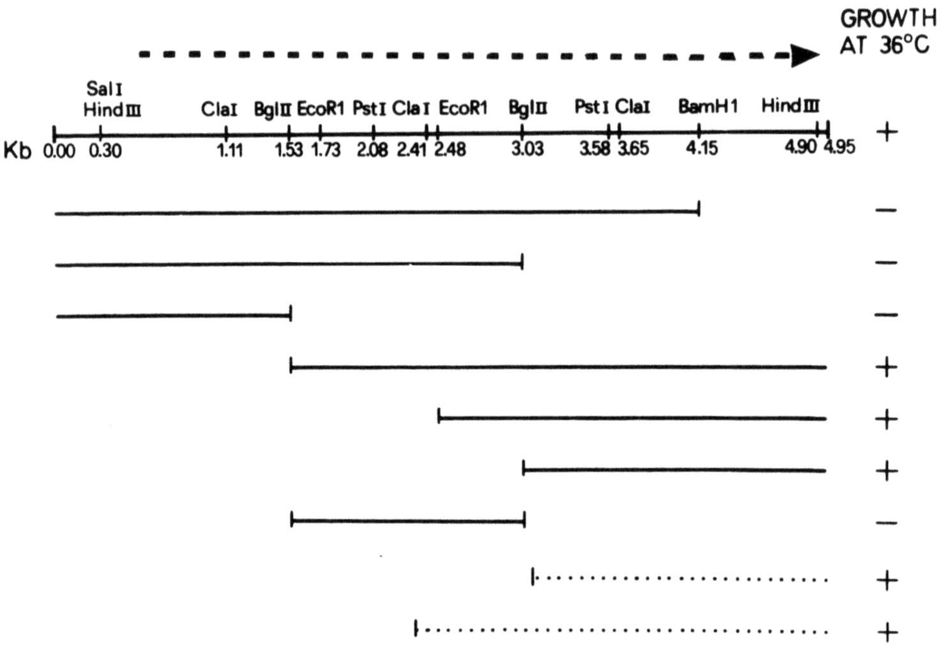

Fig. 1. Restriction map of the cloned sequence complementing *cdc25-1* mutation and subcloning results. Definite restriction endonuclease fragments *(continuous lines)* of the sequence were tested for the ability to transform TC7 strain to ts^+ phenotype. Functional selection of random generated Sau3A fragments of pDGEm-1 revealed two minor fragments still able to complement *cdc25-1* lesion *(dotted lines)*. *Dotted arrow* indicates direction and approximate initiation site of transcription

Molecular Analysis of the Start *CDC25* Gene

Isolation of the *CDC25* Gene

A yeast genomic library constructed in the *E. coli-S. cerevisiae* shuttle vector YRp7 (Nasmyth and Reed 1980) was employed to transform the recipient *S. cerevisiae cdc25-1* strain selecting for trp^+ transformants able to grow at restrictive temperature (36°C).

From some of the positive clones total nucleic acids were extracted and the plasmids were amplified in *E. coli*, and one plasmid (called pDGEm-1) was isolated; it carries a DNA insert of 5 kb and transforms TC7 strain to trp^+ and trp^+ ts^+ phenotype with the same efficiency.

A restriction map of the 5 kb DNA sequence has been obtained (Fig. 1) and subcloning experiments indicate that a plasmid carrying the 2 kb BglII-BamHI-SAU3A fragment (pDGM-4) can give complementation of the *cdc25-1* growth lesion (Fig. 1). Thus the BamHI site is most likely located within the gene sequence.

A Southern blot analysis performed on total *S. cerevisiae* DNA indicates that the cloned sequence corresponds to a unique, strongly homologous, yeast gene.

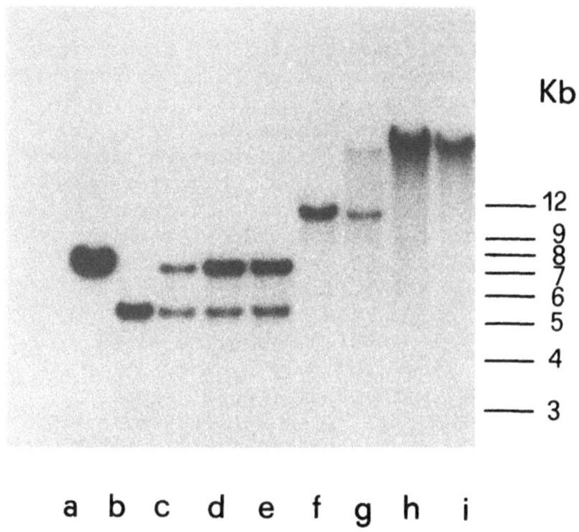

Fig. 2. Southern blot analysis of pDGEm-1 integrative transformants. BglII-BamH1-HindIII fragment of cloned sequence ^{32}P-labeled by nick-translation was used as probe. pDGEm-1 digested with EcoR1 (a); yeast total DNA of a w.t. strain (S288C) (b,f); DNA extracted from tX4 (c,g), tB12 (d,h), and tB13 (e,i). Yeast DNA's were completely digested with EcoR1 (lanes from b to e) or XhoI (lanes from f to i). No XhoI sites are present within the pDGEm-1 sequences

To establish whether the cloned sequence corresponds to the *CDC25* gene, it was integrated into the corresponding chromosomal locus and the site of integration was identified by segregation experiments after a cross of the integrative transformants with a wild-type strain. The plasmids were linearized to direct integration in the corresponding chromosomal locus (Orr-Weaver et al. 1981): pDGEm-1 was cut with BamH1 within the cloned region required for *cdc25* complementation or with XbaI in the TRP1 region.

The strain TC7 was transformed with the linearized plasmid using the LiCl method (Ito et al. 1983) which prevents cell fusion. From several *trp*$^+$*ts*$^+$ stable transformants one pDGEm-1/XbaI transformant (called tX4) and two pDGEm-1/BamHI transformants (tB12 and tB13) were further analyzed by the Southern blot technique to establish the site of plasmid integration.

As shown in Fig. 2 the cloned sequence appears to be integrated in an homologous genomic sequence in the tB12 and tB13 clones, while the putative *CDC25* genomic sequence is intact in the tX4 clone. No free plasmids are detectable in the analyzed transformants.

The transformants were then crossed with the wild-type strain 6251-48B3 and the phenotype of the spores generated by meiosis of diploids was scored. The segregation of *trp*$^-$ indicates that in the tX4 clone the integration of pDGEm-1 sequence occurred in the *TRP1* locus, while in the tB12 and tB13 clones the integration occurred in a region not linked to *TRP1*. This region of integration, which is homologous to the cloned sequence, is linked to the *CDC25* locus. We observed a low frequency (2 - 5%) segregation of the *ts*$^-$ phenotype that appears due to a mitotic or meiotic instability of the duplicated chromosomal regions of integration. In fact, Southern blot analysis on the total DNA extracts from the four spore clones of at 2ts$^+$:2ts$^-$ ascus of a tB13/6251-48B3 cross shows that the hybridization pattern is independent of the *ts* phenotype, and that the integrated plasmid sequences are no longer detectable in all these clones.

Taken together, these data indicate that most likely the integration of the pDGEm-1 sequence in tB13 and tB12 strains occurred in the *CDC25* locus and that in the pDGEm-1 plasmid it is present an yeast DNA sequence bearing the true *CDC25* gene or, at least, a functional part of this gene.

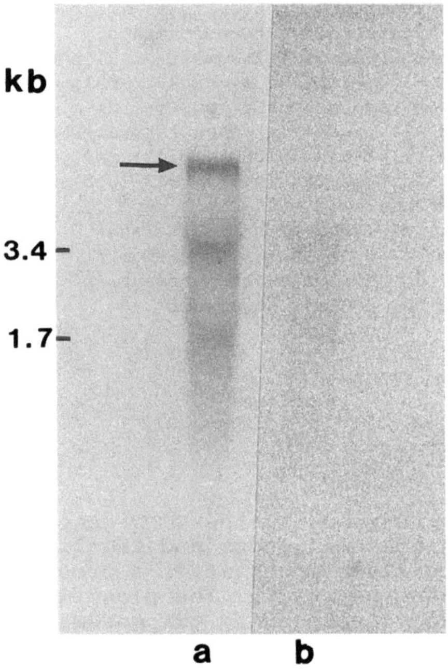

Fig. 3. Identification of *CDC25* transcript. Identical samples of about 20 µg poly$(A)^+$ RNA from wild type strain S288C separated on 1% agarose-formaldehyde denaturing gel were hybridized with complementary single-strand ^{32}P-labeled probes. These were prepared by cloning in the two different orientations the fragment EcoR1-BamH1-HindIII of the *CDC25* sequence in pSP64 (lane a) and pSP65 (lanbe b) vectors and by SP6 RNA polymerase transcription (Melton et al. 1984). A specific signal *(arrow)* is generated only by incubation of yeast RNA with pSP64 (CDC25) probe. Residual rRNA still present after oligo-dT purification step gives a specific hybridization signals on both samples

Our findings are in full agreement with the results reported by Camonis et al. (1986) that have independently cloned the *CDC25* gene.

Transcription of the *CDC25* Gene

A specific mRNA species of about 5 kb was detected by Northern blot hybridization in poly$(A)^+$ RNA extracted from wild-type S288C yeast cells. The direction of transcription appears to be SalIBamHI-HindIII (Fig. 3).

S1 mapping experiments show that the entric cloned sequence present on pDGEm-1 is protected by poly$(A)^+$ enriched RNA. This fact indicates that almost all the gene sequence is contained in the cloned fragment, which may lack regulatory sequence(s) in 5' and/or in 3'. Camonis et al. (1986) reported a 5.2 kb transcript from *CDC25* gene, and the transcription appears to initiate near the Sal1 site.

The *CDC25* gene function is required for the commitment of the yeast cells to the mitotic cycle and its inactivation brings to a G0/G1 arrest (Martegani et al. 1984, Iida and Yahara 1984a,b). The switch between alternative growth pathways could be controlled by regulating the level of *CDC25* transcripts. To test this hypothesis, the level of *CDC25* mRNA was monitored in exponential growth, in stationary phase and in the recovery of growth of starved cells. Poly$(A)^+$-enriched RNA extracted from stationary cells obtained by nitrogen starvation (Johnston 1977) and early after inoculation of starved cells in fresh, nitrogen-supplemented medium. The specific *CDC25* transcript was present in quite a low amount in all these conditions, as well as in exponentially growing cells.

These finding suggest a constitutive transcription of *CDC25* gene, or at least a transcription that is not dependent on the growth conditions.

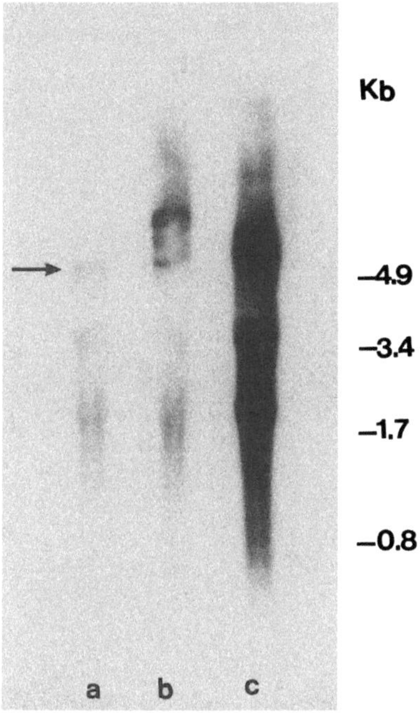

Fig. 4. *CDC25* Transcripts in transformed cells. lane a, RNA from exponential growing cells (S288C), lanes b and c, RNA extracted respectively from pDGEm-1 and pDGM-4 transformants. The amount of poly$(A)^+$ RNA was 30 µg for all the lanes. *Arrow* indicates the 5 kb *CDC25* mRNA

It would be of interest to study transcription also during differentiative pathways, in particular during sporulation, as the *CDC25* gene appears to play a role in the sporulation events. Many of the start genes so far cloned from budding yeast and from fission yeast are transcribed in a cell cycle-independent way. This finding suggests that the metabolic signals regulating the traverse of start may act by modulating the activity of the product of start genes which are constitutively synthesized. Interesting evidence along this line comes from the discovery that product of *CDC2* gene of *Schizosaccharomyces pombe* is phosphorylated in a growth-dependent way (Durkacz et al. 1986).

Several species of hybridizing mRNA were detected in pDGEm-1 and pDGM-4 transformants exponentially growing at 36°C. Besides a 5-kb RNA species, several bands at higher molecular weight are present in both strains. Moreover, in the pDGM-4 transformants, various species distributed between 2 and 3.5 kb are detected. All these different species are overexpressed in pDGEm-1 and pDGM-4-containing cells. These different RNA species present in transformed strains are specific transcripts since a complementary probe from pSP65 fails to detect them; they may be due to different transcriptional initiation sites present on the plasmids, as well as to heterogeneity in transcriptional termination sites (Fig. 4).

Search for Homologous Eukaryotic Sequences

CDC25 gene product seems to play a significant role in the regulation of growth and cell cycle progression in yeast cells (Martegani et al. 1984), and therefore it is of interest to find out whether a similar function is conserved in other eukaryotes.

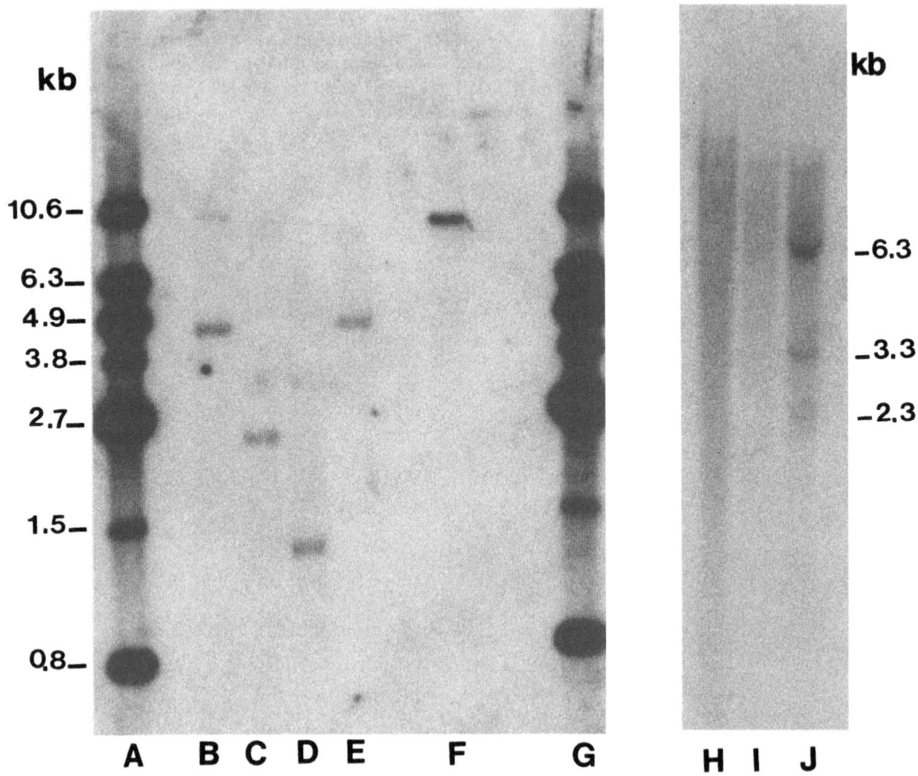

Fig. 5. Identification of sequences homologous to *CDC25* gene in DNA of different eukaryotes by Southern blot hybridization. The labeled probe (pSP64-CDC25) was the same used for Northern blot hybridization in Fig. 3. Genomic DNA's were digested with appropriate restriction endonucleases and separated on 0.7% agarose gel, transferred to nitrocellulose filter and hybridized in low-stringency conditions (35% formamide, 5 x SSPE at 42°C). Standard MW fragments (A,G); 1 µg of *S. pombe* DNA digested with HindIII (B), HindIII/EcoRI (C), HindIII/EcoRI/Pst1 (D), Pst1 (E); o.5 µg of *S. cerevisiae* DNA digested with XhoI (F); 10 µg of human DNA digested with EcoRI (H); 5 µg of mouse DNA digested with EcoRI (I); 10 µg of *N. crassa* DNA digested with EcoRI (J)

Indeed, for other cell cycle genes such homology has been detected. The class I start gene *CDC28* which codes for a phosphoprotein kinase, modified by phosphorylation, shows a significant degree of homology with the mammalian protein kinase family of oncogenes (Lorincz and Reed 1983, Reed et al. 1985). Moreover, it has an extensive homology with the *CDC2* gene of *S. pombe* and it can functionally complement the *cdc2* mutation when expressed in fission yeast (Beach et al. 1982).

Besides yeast G-like proteins that participate to the regulation of *S. cerevisiae* cell cycle by modulating the adenylate cyclase activity (Toda et al. 1985) have been identified by taking advantage of their extensive homology with the ras oncogenes.

At first we looked to detect homology between the *CDC25* sequence and several purified cloned mammalian oncogenes *(c-myc, v-myb, v-fes, c-mos, v-abl, v-scr, h-ras)* (Bishop 1985) by low stringency Southern hybridization experiments. In all cases no significant molecular hybridization was detected.

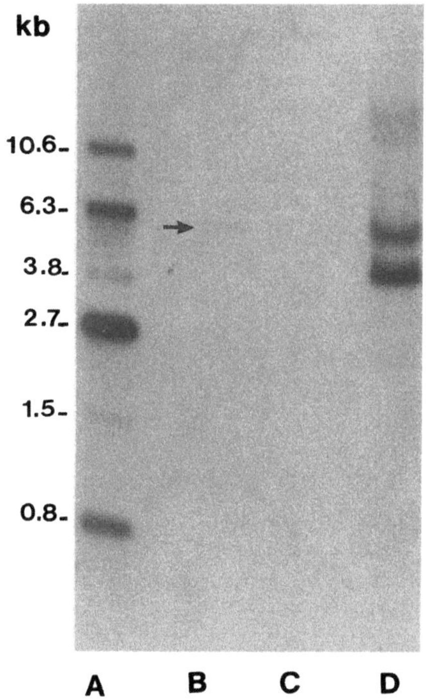

Fig. 6. Southern blot hybridization between ribosomal-DNA sequences of *S. cerevisiae* and *CDC25* gene. (A) standard MW DNA; (B) 10 ng of pRDNA6 plasmid digested with HindIII; (C) 10 ng of pMET5 plasmid digested with HindIII; (D) 5 µg of total *S. cerevisiae* DNA digested with HindIII. *Arrow* indicates the 6.4 kb fragment containing 17 and 25S RNA genes

Camonis et al. (1986) compared the sequence of the *CDC25* gene and found no significant homology between the putative amino acid sequence of the *CDC25* 170 kDa gene product and known proteins; their analysis specifically failed to reveal also functional domains for kinase activity.

Then we analyzed if DNA sequences in other eukaryotic genomes were specifically detected by a *CDC25* probe, the shorter functional fragment complementing the *ts* phenotype. Even in very low stringency conditions it was not possible to detect any specific signal of hybridization with genomic DNA of human and mouse origin. Instead, when *Neurospora crassa* or *S. pombe* total DNA (digested by various restriction enzymes) were probed, strong specific signals were detected, demonstrating that these lower eukaryotes contain sequences probably very closely related to the *CDC25* gene (Fig. 5).

We are now constructing the restriction map of such hybridizing fragments. The cloning of these *CDC25*-like sequences from *N. crassa* and *S. pombe* and their structural and functional analysis are under way. Particularly interesting is the possibility of expressing them in *S. cerevisiae* and studying the phenotypic modifications of the *cdc25* cells.

The *CDC25* Gene Shows a Low Homology with Yeast Ribosomal DNA Sequences

During molecular hybridization studies, partial signals of hybridization between high specific activity *CDC25* probes and both 17S and 25S *S. cerevisiae* ribosomal RNA have been observed, even when poly(A)$^+$-enriched RNA samples have been used instead of total RNA and when the experiments were performed in high stringency conditions of hybridization (50° - 60°C, 50% formamide, 5 x SSC; Fig. 3).

The specificity of the ribosomal DNA sequences in this hybridization with the *CDC25* probe is supported by a Southern blot analysis performed by transferring onto nitrocellulose filter a purified pDB248 recombinant plasmid bearing a 6.4 kb HindIII yeast DNA fragment that contains the genes for 17S and 25S rRNA's, together with the total yeast DNA digested with HindIII and a pMET5 plasmid that contains the *S. cerevisiae* MET2 gene cloned in the pDB248 (Baroni et al. 1984). As shown in Fig. 6, when this filter is hybridized with the EcoR1-BamH1-HindIII labeled fragment of *CDC25* gene at 42°C in 35% formamide, a signal was detected in both cloned rDNA and in the corresponding 6.4 kb sequence in total yeast DNA. Instead no hybridization was observed with the MET2 sequence.

The extension and significance of this homology is presently under investigation.

Physiological Role of the *CDC25* Gene

Overexpression of the *CDC25* Gene and Alteration of the Cell Cycle Progression in Yeast Cells

The *cdc25-1* strain, after a shift to restrictive temperature, undergoes structural and physiological modifications typically associated with the starvation condition of the yeast cells (Martegani et al. 1984). In fact after the shift of a *cdc25-1* culture from 24° to 36°C, a rapid and severe inhibition of the stable RNA synthesis can be observed, while there is a limited accumulation of proteins. After 4 - 6 h they become cells specifically arrested in the G1-unbudded phase of the cell cycle, acquire high levels of thermotolerance, accumulate large quantities of storage carbohydrates and their cell wall becomes much more resistant to lythic enzymes, thus resembling cells entering in stationary phase (Martegani et al. 1984, 1986).

Lesions of the permeability of nutrients or of the intermediate metabolism are unlikely and moreover the fact that homozygous *cdc25* a/α diploids can sporulate at 36°C (Shilo et al. 1978) indicates a substantial integrity of the transcriptional and translational apparatus also under restrictive conditions.

Considering that the *cdc25-1* mutant is derepressed for sporulation (Shilo et al. 1978) and that it shows a "GO-like" phenotype at restrictive temperature, we made the hypothesis that the *CDC25* gene product may have a role in a regulative mechanism which leads the cells to choose between the entrance in the cell division cycle and the beginning of differentiative pathways related to nutrient depletion (Martegani et al. 1984).

Such a hypothesis has now been tested by utilizing the cloned sequence of the *CDC25* gene.

For this purpose we have investigated whether the presence of the *CDC25* gene on a multicopy plasmid modifies the control of the cell cycle in budding yeast, and in particular the ability to arrest the cell cycle under stationary phase conditions. While control clones, transformed with YRp7, show a marked decrease of the percentage of budded cells in stationary phase of growth (no increase in cell number for 48 h), a high percentage of budded cells was found in pDGEm-1 and pDGM-4 transformants (Table 1). This finding suggests that the *CDC25* gene on a multicopy plasmid and its overexpression, which has been detected as specific transcripts (Fig. 4), may cause an alteration in the cell cycle

Table 1. Fraction of budded cells in stationary phase cultures of pDGEm-1 transformants

Strain	24°C		36°C	
	E	S	E	S
A-TC7/pDGEm-1	0.50	0.35	0.51	0.30
A-TC7/YRp7	0.52	0.21	-	-
D-TC7/pDGEm-1	0.58	0.48	0.55	0.50
D-TC7/YRp7	0.58	0.12	-	-
A-TC7/pDGM-4	0.60	0.30	0.58	0.42
tB12	0.60	0.42	0.58	0.38
TX4	0.54	0.43	0.56	0.38
TY7	0.52	0.19	-	-
TC7	0.50	0.20	-	-
S288C	0.65	0.20	0.59	0.11

[a]Cultures of transformants were grown in YNB/Glucose medium without tryptophane. The % of budded cells was determined during exponential (E) and late stationary phase (S)
A-TC7: haploid transformants; D-TC7: diploid transformants

progression of transformed cells. Moreover, this alteration is more evident in diploid clones than in haploid ones, and this finding has been related to the stability of the plasmids. In fact, at the end of the growth experiment, only 15 - 20% of the haploid cells retain the plasmids at the end of growth as against 50 - 70% of diploid cells. Also integrative transformants showed a similar alteration in the cell cycle control. In fact, while the strain TY7, obtained by transforming TC7 with YRp7 linearized with XbaI, shows a normal response to stationary phase conditions, both tB12, tB13, and tX4 strains fail to arrest in the G1 unbudded phase of the cell cycle after 2 days in stationary phase, as demonstrated also by flow-cytometric analysis of DNA distributions. Furthermore, we have found 60% viable cells in stationary phase cultures of TY7, while only 2 and 8% in tB12 and, respectively, in tX4 strains. This alteration of cell cycle regulation may again be related to the altered expression of the integrated *CDC25* sequences that in all three transformants are flanked by plasmid sequences. A densitometric analysis performed on the Southern blot of Fig. 3 shows that five to six copies of pDGEm-1 are integrated at the *CDC25* locus in tB12 and tB13 strains, and that only one copy is integrated in the TRP1 locus in TX4 strain, but obviously the *CDC25* sequence is flanked by plasmid sequences on both sides.

Taken together, these findings suggest that the product of the *CDC25* gene is a positive regulator of the cell cycle progression and that its activity may be modulated by signals related to the nutritional and to the metabolic conditions of the yeast cells.

To obtain a more complete picture of the physiological role of the *CDC25* gene, we are presently investigating the regulation of the sporulation in diploid a/α transformed with plasmids bearing the *CDC25* gene.

The *CDC25* Gene Product Is Involved in a Regulatory Mechanism Activated by cAMP

The relevance of cAMP for the cell cycle regulation and for the coordination of differentiative pathways in budding yeast was ascertained when mutants bearing defined lesions in the cAMP metabolism were isolated and their phenotype analyzed.

Yeast cells with lesions in adenylate cyclase (*cyr1* mutant; Matsumoto et al. 1983a,b) or its positive regulator RAS2 (Toda et al. 1985, Tatchell et al. 1985) have a reduced intracellular level of cAMP and show a class II start terminal phenotype. In other words, they mimic, just like the *cdc25* mutant, stationary phase cells and they are derepressed for sporulation if a/α diploids. The same terminal phenotype is caused by lesions in the regulatory *(CYR3)* or in the catalytic *(cyr2)* subunit of cAMP-dependent protein kinase (cA-pK) (Matsumoto et al. 1985a)

All the above-mentioned mutants are temperature-sensitive or need exogenous cAMP for growth. Extrallelic supressors of the *cyr* or *ras* mutations which restore growth in restrictive conditions or without the addition of exogenous cAMP, identify mutations leading to the stimulation of cAMP production (RAS^{Val19}, *IAC*) or to constitutive high levels of a cA-pK activity independent of cAMP *(byc1)* (Toda et al. 1985, Matsumoto et al. 1985a). When limited for nutrients, these revertant clones show a defective cell cycle, being unable to preferentially arrest in G1, as wild-type cells do; besides they are not able to accumulate storage carbohydrates and rapidly lose viability.

These findings suggest that cAMP works as a positive factor in promoting cell cycle progression from Start. cAMP-dependent protein kinase is believed to mediate the effects of cAMP in a cascade that leads to phosphorlyation of presently unknown target proteins as response to the production of the regulatory cyclic nucleotide. Another goal is to clarify how the nutritional signals are transduced and how they can modulate RAS proteins and adenylate cyclase activities.

On the other hand, CDC genes have been found to be involved in cAMP metabolism. The Start *CDC35* gene was found to be allelic to *CYR1* and to codify for adenylate cyclase (Casperson et al. 1985, Boutelet et al. 1985, Kataoka et al. 1985) and functional relationships have been shown between the *CDC25* gene product and cAMP metabolism. Mutations in the *CDC25* sequence lower the cAMP level (in the *cdc25-5* mutant; Camonis et al. 1986) and/or cAMP-dependent protein kinase activity (in the *cdc25-1* mutant; our preliminary experiments) apparently in allelic-specfic manner.

Addition of exogenous cAMP to *cdc25-1* cultures shifted to restrictive temperature causes a partial reversion with an increase of the fraction of budded cells and a stimulation of nuclear DNA synthesis (Martegani et al. 1986). Moreover, a double mutant *cas1 cdc25-5* grows well if the culture medium is supplemented with cAMP (Camonis et al. 1986, F. Hilger personal communication).

The gene that codes for a putative cAMP-dependent protein kinase has been cloned as an extrallelic suppressor of the *cdc25* mutation (T. Toda personal communication).

Finally, revertants of *ras2* mutants carrying sra mutation which have elevated cAMP levels or cA-pK activity, suppress the *ts* phenotype of *cdc25-1* (K. Tatchell personal communication).

Although fragementary, data indicate that an impairment of the *CDC25* gene function may lead to inactivation of the cAMP production. In turn, an enhanced cAMP production may correct growth defects derived by the *CDC25* lesion.

The *CDC25* gene product may then act as a positive regulator of the cAMP production. The determination of cAMP and of cA-pK activity in cells in which overexpression of *CDC25* functional sequence can be induced may be informative.

Because a short portion of the cloned *CDC25* gene is still able to complement the growth defect of mutant *cdc25-1* cells (see subcloning data in Fig. 1) we suppose that the large *CDC25* protein (about 170 kDa of molecular weight) has probably a multidomain structure.

A multifunctional *CDC25* gene product might participate in integrative processes between different regulative pathways that operate at Start in response to various signals about cellular conditions. To locate the role of this gene in a more complete network of functional interelationships, we isolated a large collection of independent revertant clones of *cdc25-1* mutant able to growth at restrictive temperature. In principle, secondary mutations that are present in such strains may identify new genes whose products can have functional or structural interaction with *CDC25* protein.

A preliminary characterization showed the existence of different phenotypic classes between revertants. It is interesting to note that some of them lose viability when starved for nutrients being unable to arrest in the stationary phase and to accumulate storage charboidrates. Thus they resemble *bcy1* or RAS^{Val19} mutants or mimic partially phenotypic modifications of transformants overexpressing *CDC25* genes sequences. Genetic and biochemical tests are necessary to establish an involvment of cAMP metabolism activation in these *cdc25-1* revertants and to identify new genes eventually operating in different pathways together with *CDC25* gene.

A Labile Protein Is Synthesized During the Recovery from a *cdc25* Arrest

All of the Start genes of *S. cerevisiae* isolated until now show a constitutive transcription; however, specific gene product(s) can be made during the Start process and therefore its identification can be relevant for the understanding of the molecular mechanism of the cell cycle control.

Kinetic evidence obtained during a release of temperature block in *cdc25-1* strains (Shilo et al. 1979, Popolo and Alberghina 1984) suggested the requirement of labile protein(s) for cell cycle initiation.

A recent analysis of the synthesis and turnover of *S. cerevisiae* proteins during the cell cycle indicates that several proteins are cycle-modulated and that some of them are unstable (Lorincz et al. 1982).

Work done in our laboratory showed the presence of at least one labile protein synthesized specifically during a release from a temperature block in *cdc25* mutant cells (Popolo and Alberghina 1984). This is a 100 kDa polypeptide (p100) with an isoelectric point of about 4.8 - 5. This protein is most likely a plasma membrane protein, and experiments performed both with tunicamycin and peptide mapping show that this protein is a glycoprotein and appear to be related with a more stable glycoprotein (p115) that is also cell cycle-regulated (Popolo et al. (1986).

Conclusions

In the preceding sections, we have shown how some questions about yeast cell cycle regulation can now be investigated at a molecular level by a combined genetic, physiological, and biochemical analysis. Such an approach has been used in this paper to clarify the physiological and molecular role of the *CDC25* gene.

Much evidence supports the idea that cAMP may be one of the signals that trigger the early stages of yeast cell proliferation and that the *CDC25* gene plays some function in this pathway.

Considering the terminal phenotypes of mutants whose defective function is related to Start, two predominant phenotypes are apparent: repressed mutants, that are no longer able to proliferate, and completely derepressed mutants that fail to arrest the cell cycle under unfavorable conditions.

While most of these mutants clearly show a defect in the cAMP metabolism or in the cAMP-protein kinase pathway *(cdc25, cdc35, cyr1, cyr2, CYR3, ras2,* RAS^{Val19}*, IAC, pde1, ppd1, bcy1)* (Matsumoto et al. 1985a,b) for others like *ard1* (Whiteway and Szostak 1985), *hsr1, suh1* (Iida and Yahara 1984c, H. Iida personal communication), and *whi1* (Sudbery et al. 1980), no clear relationship with the cAMP metabolism has been demonstrated so far, suggesting the existence of multiple regulatory levels at Start.

Acknowledgment. We thank Dr. I. Biunno for the gift of cloned oncogenes sequences and Prof. S. Sora for the helpful discussion of genetic experiments. This work was supported by C.N.R. grant-Target Project: Ingegneria Genetica e Basi Molecolari delle Malattie Ereditarie, No. CT 84.00861.51.

References

Baroni M, Velati-Bellini A, Vanoni M, Martegani E (1984) Abstr SIB, Ischia, Italy
Beach D, Durkacz B, Nurse P (1982) Nature (London) 300:706-708
Bishop JM (1985) Cell 42:33-38
Boutelet F, Petitjean A, Hilger F (1985) EMBO J 4:2635-2641
Camonis JH, Kalekine M, Gondrê B, Garreau H, Boy-Marcotte E, Jacquet M (1986) EMBO J 5:375-380
Casperson GF, Walker N, Bourne HR (1985) Proc Natl Acad Sci USA 82:5060-5063
Dawes IW, Calvert GR (1984) J Gen Microbiol 130:605-613
Durkacz B, Carr A, Nurse P (1986) EMBO J 5:369-374
Hartwell LH (1974) Bacteriol Rev 38:164-198
Hartwell LH, Mortimer LK, Culotti J, Culotti M (1973) Genetics 74:267-286
Hartwell LH, Culotti J, Pringle JR, Reid BJ (1974) Science 183:46-51
Hinnen A, Hicks JB, Fink GR (1978) Proc Natl Acad Sci USA 75:1929-1933
Iida H, Yahara I (1984a) J Cell Biol 99:199-207
Iida H, Yahara I (1984b) J Cell Biol 98:1185-1193
Iida H, Yahara I (1984c) J Cell Biol 99:1441-1450
Ito H, Fukuda Y, Murata K, Kimura A (1983) J Bacteriol 153:163-168
Johnston GC (1977) J Bacteriol 132:738-739
Kataoka T, Broek D, Wigler M (1985) Cell 43:493-505
Lorincz AT, Reed SI (1983) Nature (London) 307:183-185
Lorincz AT, Miller MJ, Xuong NH, Geiduschek EP (1982) Mol Cell Biol 2:1532-1545
Martegani E, Vanoni M, Baroni M (1984) Eur J Biochem 144:205-210
Martegani E, Baroni M, Vanoni M (1986) Exp Cell Res 162:544-548

Matsumoto K, Uno I, Ishikawa T (1983a) Exp Cell Res 146:151-161
Matsumoto K, Uno I, Ishikawa T (1983b) Cell 32:417-423
Matsumoto K, Uno I, Ishikawa T (1985a) Yeast 1:15-24
Matsumoto K, Uno I, Kato K, Ishikawa T (1985b) Yeast 1:25-38
Melton DA, Krieg PA, Rebagliati MR, Maniatis T, Zinn K, Green MR (1984) Nucleic Acids Res 18:7035-7056
Nasmyth KA, Reed SI (1980) Proc Natl Acad Sci USA 77:2119-2123
Nurse P (1981) In: Gull K (ed) The fungal nucleus. Cambridge Univ Press, New York, pp 330-345
Orr-Weaver TL, Szostak JW, Rothstein RJ (1981) Proc Natl Acad Sci USA 78:6354-6358
Popolo L, Alberghina L (1984) Proc Natl Acad Sci USA 81:120-124
Popolo L, Vai M, Alberghina L (1986) J Biol Chem 261:3479-3481
Pringle JR, Hartwell LH (1981) In: Strathern GN, Jones EW, Broach JR (eds) The molecular biology of the yeast Saccharomyces cerevisiae-life cycle and inheritance. Cold Spring Harbor Lab, Cold Spring Harbor New York, pp 97-142
Reed S (1980) Genetics 95:561-577
Reed SI, Ferguson J, Groppe GC (1982) Mol Cell Biol 2:412-425
Reed SI, Hadwiger JA, Lorincz AT (1985) Proc Natl Acad Sci USA 82:4055-4059
Shilo V, Simchem G, Shilo B (1978) Exp Cell Res 112:241-248
Shilo B, Riddle VGH, Pardee AB (1979) Exp Cell Res 123:221-227
Sudbery PE, Goodey AR, Carter BLA (1980) Nature (London) 288:401-403
Tatchell K, Robinson LC, Breitenbach M (1985) Proc Natl Acad Sci USA 82:3785-3789
Toda T, Uno I, Ishikawa T, Powers S, Kataoka T, Broek D, Cameron S, Broach J, Matsumoto K, Wigler M (1985) Cell 40:27-36
Whiteway M, Szostak JW (1985) Cell 43:483-492

Regulation of Histone Gene Transcription in *S. cerevisiae*

M. A. Osley[1], J. Gould[2], S. Kim[3], D. Lycan[1], and L. Hereford[1]

Introduction

The histone genes belong to a subset of eukayrotic genes whose expression is temporally regulated during the cell cycle. Histone RNAs accumulate only during S phase, through the combination of an elevated rate of histone gene transcription and increased histone mRNA stabilization (Detke et al. 1979, Hereford et al. 1981, 1982, Heintz et al. 1983, Sittman et al. 1983, Artishevsky et al. 1984). The regulation of transcription is the first, if not the major point of control in the synthesis of histone proteins, because in the normal cell cycle histone mRNA synthesis significantly precedes the accumulation of these mRNAs (Hereford et al. 1982, Plumb et al. 1982, Artishevsky et al. 1984).

Using *S. cerevisiae* histone genes, we have obtained information in three general areas which is pertinent to the problem of transcriptional control. First, we have determined that the temporal restriction of histone gene transcription is dependent upon completion of steps in the cell cycle pathway. Second, we have identified *cis*-acting sequences in the histone genes which regulate transcription. Third, we have initiated studies to identify gene products which interact with transcriptional control sequences.

Histone Gene Transcription Is Dependent on the Cell Cycle

In *S. cerevisiae* the histone genes are transcribed during late G1-early S phase (Hereford et al. 1982). Two possibilities can be considered for the temporal restriction of transcription during the cell cycle. Transcription could be timed by a cellular "clock" which runs simultaneously with, but independently of, steps in the cell cycle pathway. Alternatively, transcription could be dependent upon completion of steps in this pathway which are themselves temporally regulated. The existence of well characterized *S. cerevisiae* mutants which block specific points in the cell cycle (Hartwell 1974) allowed us to distinguish between these possibilities (Hereford et al. 1982). These studies showed that initiation of histone gene transcription is dependent upon completion of the G1 step called Start, which is an important control point in the yeast cell cycle. Turn-off transcription is dependent upon completion of a later step in G1 which is defined by the *cdc7* mutation and is required for initiation of DNA replication.

[1]Dana Farber Cancer Institute, 44 Binney Street, and Department of Microbiology and Molecular Genetics, Harvard Medical School, Boston, MA 02115 USA

[2]Department of Biological Sciences, University of Warwick, Coventry, West Midlands, England

[3]Department of Biochemistry, Oxford University, South Parks Road, Oxford, England

37. Colloquium Mosbach 1986
Cell Cycle and Oncogenes

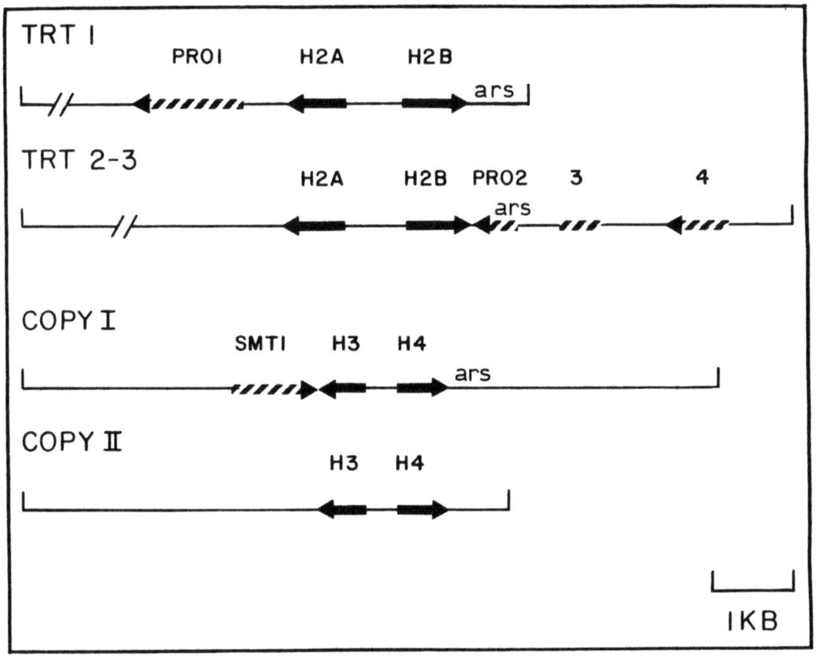

Fig. 1. Organization of histone gene loci in *S. cerevisiae*. The four core histones of *S. cerevisiae* are encoded by four sets of unlinked gene pairs (Hereford et al. 1979, Smith and Andresson 1983). The *H2A-H2B* gene set defined by *TRT1* resides on chromosome IV, closely linked to *rad9*. The second *H2A-H2B* gene pair *(TRT2-3)* maps to chromosome II in the order *TRT2-3-CEN2-gal1*. The *CopyI H3-H4* gene pair resides on chromosome IX in the order *his6-CEN9-CopyI-lys1*. The remaining *H3-H4* gene set (CopyII) is unlinked to its centromere or to the other histone gene pairs

Cis-Acting Sequences Which Regulate the Timing of Transcription

The four unlinked histone gene loci in yeast are characterized by a common structural organization, in which approximately 1 kb of spacer DNA separates a divergently transcribed *H2A-H2B* or *H3-H4* gene pair (Fig. 1). In addition, in three of the four loci, *ars* elements are found in close proximity to the 3' end of the *H2B* or *H4* gene. Since these elements are believed to represent origins of chromosomal DNA replication (Chan and Tye 1980), we have proposed that in this particular capacity they also regulate the timing of histone gene transcription (Osley and Hereford 1982). Thus, the restriction of transcription to the period in the cell cycle around the initiation of DNA replication could be accounted for by periodic activation of the histone gene-associated *ars* elements. We have recently shown, however, that the *ars* is dispensible as a transcription timer, because a *H2A-lacZ* fusion gene is transcribed with correct cell-cycle periodicity whether or not the *ars* is present (Fig. 2). Since the *ars* does appear to be required to achieve maximal levels of *H2A-lacZ* RNA, we cannot eliminate the possibility that this element has another role in cell cycle-dependent expression. This question is being examined in more detail by replacing a chromosomal *ars* at one of the histone loci with mutant derivatives.

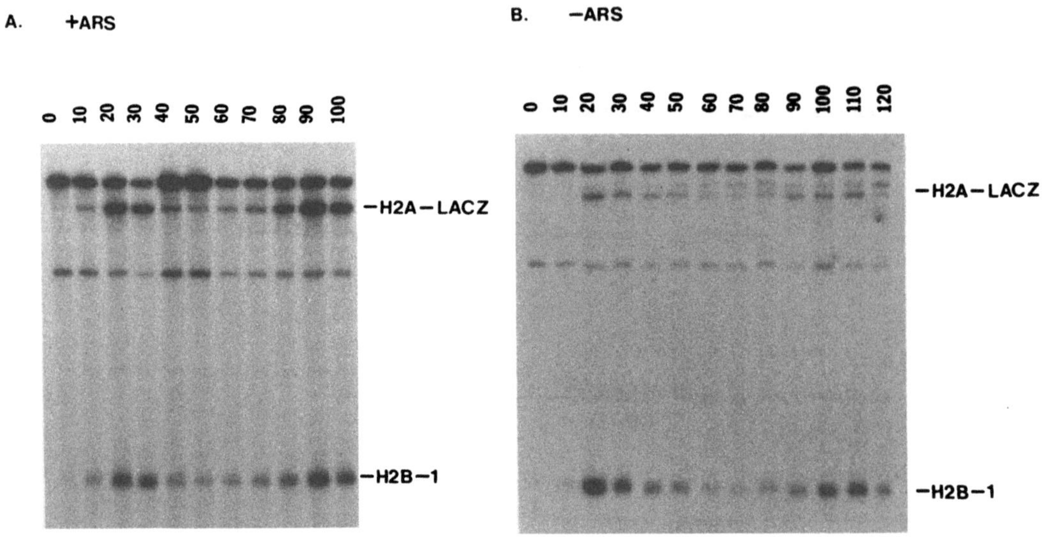

Fig. 2. Effect of the *TRT1 ars* on the timing of histone gene transcription. An *H2A-lacZ* fusion gene was integrated at *TRT1* (A), so that it was ajacent to the chromosomal *ars* at that locus, or at *leu2* (B), which has no nearby *ars*. Transcription of the *H2A-lacZ* gene and the chromosomal *H2B1* gene was followed during the synchronous cell cycles using a S1 nuclease protection assay. The additional bands represent reannealed probe

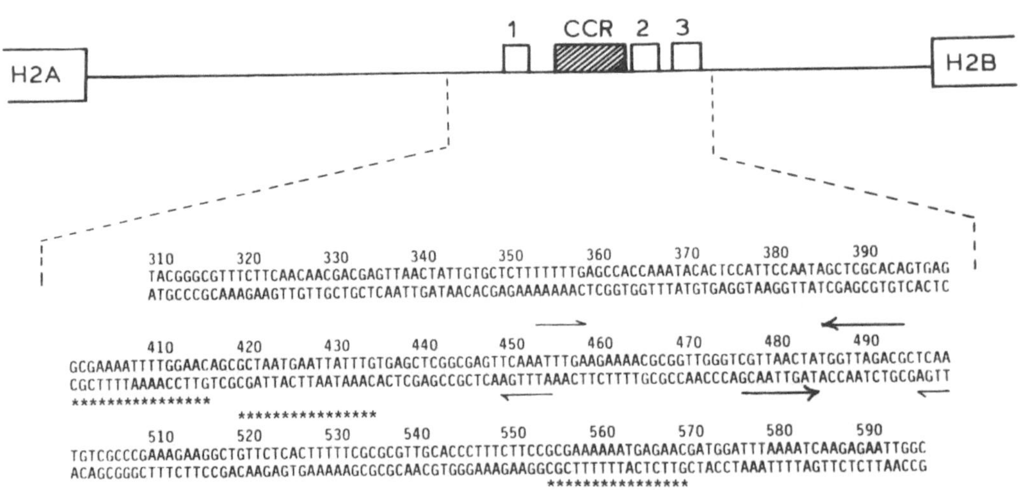

Fig. 3. Organization of regulatory elements in the *TRT1* upstream promoter. The nucleotide sequence of the *TRT1* intergene region between positions 310 and 597 is shown. This region represents the maximal extent of essential upstream promoter sequences, because when it is deleted, neither *H2A* nor *H2B* RNA is made. The open boxes represent three 16 bp UAS elements, which are indicated on the nucleotide sequence with asterices. *Hatched box* represents a region containing negative regulatory sequences (CCR). *Thick arrows* indicate a sequence of dyad symmetry in the CCR, and *thinner arrows* indicate half-site sequences from this dyad

The elimination of the *ars* as a transcriptional timing element meant that timing controls reside in 5' promoter sequences. To identify cell cycle-dependent regulatory elements in these sequences, we used two complementary experimental approaches (Osley et al. 1986). Deletions were made in vitro in promoter sequences and analyzed for their effects on expression and cell cycle regulation by fusion to the bacterial *lacZ* gene present on a yeast centromere vector. Regulatory elements identified by this method were then inserted into a heterologous yeast promoter to study their effects in isolation.

These two analyses have shown that the sequences separating the *H2A-H2B* genes at the *TRT1* locus contain a bifunctional upstream promoter of approximately 150 bp (Fig. 3). Two separable transcriptional regulatory elements are found in this region. Gene activation is achieved by a repetitious 16 bp sequence which is formally equivalent to a UAS (upstream activating site; Struhl 1982, Guarente et al. 1982, Guarente and Mason 1983, Guarente 1984) element on the basis of two criteria. First, the level of histone gene transcription is roughly proportional to the number of these repeats. Second, isolated elements can substitute for the heterologous *CYC1* UAS to activate a *CYC1-lacZ* (Guarente and Ptashne 1981) fusion gene. This UAS may constitute a histone gene-specific UAS because it is found in two additional yeast histone promoters. One copy of the entire 16 bp consensus sequence is found in the *CopyI H3-H4* promoter, and three copies of a conserved core sequence from the element (GCGAAA) are located in the *CopyII H3-H4* promoter (Smith and Andresson 1983). Although the promoter of *CDC9*, another cell cycle-regulated yeast gene (Peterson et al. 1985), does not contain the histone UAS sequence (L. Johnson personal communication), the promoter of a second periodic gene, *HO* (Nasmyth 1983) contains a UAS with the motif GACGAAAA (Nasmyth 1985).

The histone UAS element is additionally distinguished by the observation that it is regulated during the cell cycle. This attribute is illustrated by its ability in isolation to periodically activate transcription of a *CYC1-lacZ* gene, so that *CYC1-lacZ* RNA is made at the same time in the cell cycle as *H2B* RNA. Since the *HO* UAS, which is partially homologous to the histone UAS core sequence, has an identical effect in a heterologous promoter (Breeden and Nasmyth 1986), it is likely that these two cell cycle-specific elements are regulated by a common mechanism.

Although the UAS appears to be sufficient to account for periodic transcription of the histone genes, it is not the only function related to cell cycle control in the *TRT1* promoter. A second function (CCR, for cell cycle regulation) contains negative regulatory sequences which periodically repress tanscription during the cell cycle. When these sequences are deleted from the histone promoter, transcription is constitutive. Unlike the UAS elements, the negative control site cannot activate transcription, but it can bring a heterologous, constitutive UAS under cell cycle control. Its effects on periodic transcription can be further differentiated from those of the UAS elements because its presence in a heterologous promoter results in an earlier period of transcription during the cell cycle relative to a *H2B* gene. A 17 bp sequence of dyad symmetry in the negative site confers cell cycle-dependent transcription, and it can function either upstream of a UAS or between a UAS and TATA box. Interestingly, the dyad may also be responsible for generally represssing transcription: its deletion from the histone promoter results in elevated levels of histone-*lacZ* transcripts and it reduces the expressed levels of *CYC1-lacZ* RNA when it is adjacent to the *CYC1* UAS.

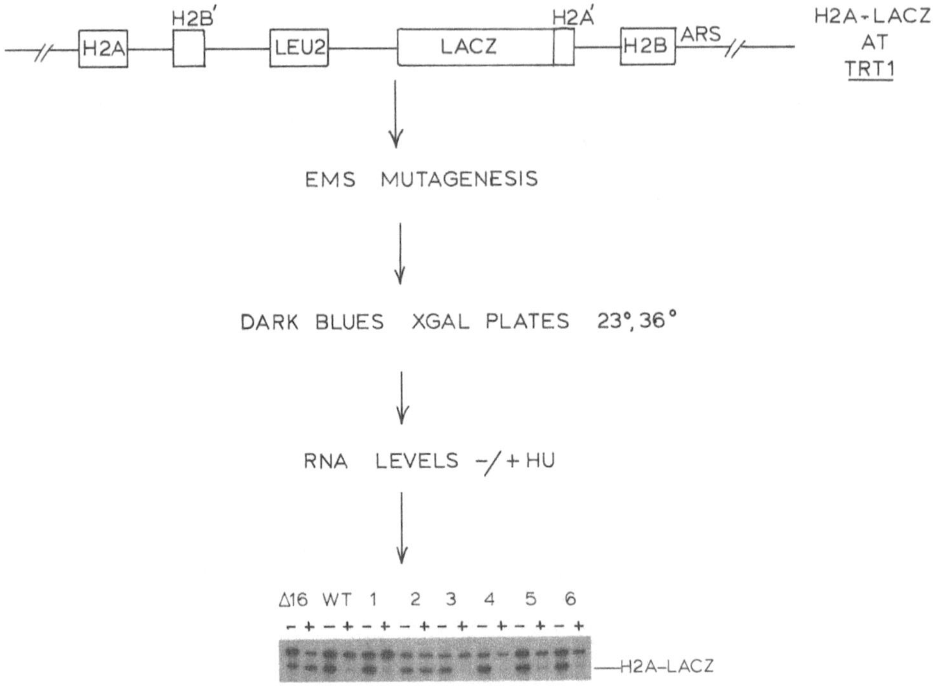

Fig. 4. Screen for trans-acting regulatory mutants. A strain containing a single copy of an *H2A-lacZ* fusion gene integrated at the *TRT1* locus was mutagenized with EMS. Surivors were replica plated to minimal media plates containing XGal and incubated at 23° and 36°C. Dark blue colonies were selected for analysis of cell cycle regulation. Putative mutants were grown exponentially (-) or arrested in S-phase (+) with hydroxyurea (HU), and the levels of *H2A-lacZ* RNA were measured by a S1 nuclease protection assay. Mutants which have lost cell cycle regulation will continue to make fusion RNA when they are blocked in S, a phenotype which is seen when the cis-acting CCR sequences have been deleted (Δ16).

Are there additional cell cycle-dependent elements in this histone promoter? The existence of another element could account for the apparently paradoxical observation that deletion of the negative control site from the *TRT1* promoter leads to cell cycle-constitutive transcription, obscuring the cell cycle-dependent activation shown by the remaining UAS elements in isolation. Our deletion analysis has so far failed to reveal the presence of other regulatory elements, and we have also eliminated the possibility that one of the three UAS sequences is itself a constitutive element. We therefore favor the view that in the *TRT1* promoter the negative control site is epistatic to the UAS elements in cell cycle regulation. Thus, we consider repression of transcription to be the predominant mechanism regulating cell cycle-dependent histone gene expression.

Identification of Trans-Acting Regulatory Proteins

Since the two classes of cell cycle elements occur in close proximity in the *TRT1* promoter, their topological arrangement may be important and thereby account for the epistatic effect of the negative control site. To determine how the two kinds of sites interact, we are attempting to isolate mutations in genes that code for cell-cycle specific regulatory proteins (Fig. 4). We assume that each cell cycle element binds a specific regulatory protein, as has been demonstrated for both positive (Giniger et al. 1985, Hope and Struhl 1985) and negative (Johnson and Herskowitz 1985) regulatory elements in several yeast promoters. To isolate mutations in genes specifying proteins that interact with the negative control sequences, we have taken advantage of our observation that deletion of this site from an *H2A-lacZ* fusion gene leads to elevated expression of β-galactosidase as measured on X-gal indicator plates. Thus, we have reasoned that a mutation in a CCR regulatory protein should produce a similar phenotype. Following EMS mutagenesis of a strain containing a single integrated copy of a *TRT1 H2A-lacZ* gene, we have screened for dark blue colonies at 23 and 36°C. The majority of cells with this phenotype are dark blue at both 23° and 37°C, while a smaller number have a temperature-sensitive response, showing elevated β-galactosidase levels only at 36°C. All of these putative regulatory mutants have been further screened for their effects on cell cycle regulation. Approximately one out of ten of the "overexpressors" has lost cell cycle regulation, indicating the efficacy of our phenotypic screen. We are currently examining these strains in more detail to determine whether the mutations are specific to the *TRT1* histone genes or more generally affect cell cycle regulation of all the histone loci and other periodically transcribed genes.

Summary

The proper timing of histone gene transcription during the cell cycle is dependent upon regulatory elements which occur in 5' promoter sequences. One class of repetitive elements is able to periodically activate transcription and is analogous to a UAS (upstream activating site) sequence. A second element has properties of a yeast operator because it is comprised of negative regulatory sequences which periodically repress transcription. The interaction between these two elements appears to be complex, and the negative element is epistatic in its effects on cell cycle regulation. The properties of the negative control sequences have been utilized to screen for trans-acting mutants in cell cycle regulation, and several potential regulatory mutants have been isolated.

Acknowledgments. This work was supported by NIH Grant GM31311 to LMH.

References

Artishevsky A, Delegeane AM, Lee AS (1984) Use of a cell cycle mutant to delineate the critical period for the control of histone mRNA in the mammalian cell cycle. Mol Cell Biol 4:2364-2369

Breeden L, Nasmyth K (1986) Regulation of the yeast *HO* gene. Cold Spring Harbor Symp Quant Biol (in press)

Chan CSM, Tye B-K (1980) Autonomously replicating sequences in *Saccharomyces cerevisiae*. Proc Natl Acad Sci USA 77:6329-6333

Detke S, Lichter A, Phillips I, Stein J, Stein G (1979) Reassessment of histone gene expression during the cell cycle in human cells by using homologous H4 histone cDNA. Proc Natl Acad Sci USA 76:4995-4999

Giniger E, Varnum SM, Ptashne M (1985) Specific DNA binding of *GAL4*, a positive regulatory protein of yeast. Cell 40:767-774
Guarente L (1984) Yeast promoters; positive and negative elements. Cell 36:799-800
Guarente L, Mason T (1983) Heme regulates transcription of the *CYC1* gene of *S. cerevisiae* via an upstream activation site. Cell 32:1279-1286
Guarente L, Ptashne M (1981) Fusion of *Escherichia coli* lacZ gene to the cytochrome C gene of *Saccharomyces cerevisiae*. Proc Natl Acad Sci USA 78:2199-2203
Guarente L, Yocum R, Gifford P (1982) A *GALIO-CYC1* hybrid yeast promoter identifies the *GAL4* regulatory region as an upstream site. Proc Natl Acad Sci USA 79:7410-7414
Guarente L, LaLonde B, Gifford P, Alani E (1984) Distinctly regulated tandem upstream activation sites mediate catabolite repression of the *CYC1* gene of *S. cerevisiae*. Cell 36:503-511
Hartwell LH (1974) *Saccharomyces cerevisiae* cell cycle. Bacteriol Rev 38:164-198
Heintz N, Sive LH, Roeder RG (1983) Regulation of human histone gene expression: kinetics of accumulation and changes in the rate of synthesis and in the half-lives of individual histone mRNAs during the HeLa cell cycle. Mol Cell Biol 3: 539-550
Hereford LM, Hartwell LH (1974) Sequential gene function in the initiation of *Saccharomyces cerevisiae* DNA synthesis. J Mol Biol 84:445-461
Hereford L, Fahrner K, Woolford T, Rosbash M, Kaback DB (1979) Isolation of yeast histone genes H2A and H2B. Cell 18:1261-1271
Hereford LM, Osley MA, Ludwig JR, McLaughlin C (1981) Cell-cycle regulation of yeast histone mRNA. Cell 24:367-375
Hereford L, Bromley S, Osley MA (1982) Periodic transcription of yeast histone genes. Cell 30:305-310
Hope I, Struhl K (1985) *GCN4* protein synthesized in vitro, binds *HIS3* regulatory sequences: implications for general control of amino acid biosynthetic genes in yeast. Cell 43:177-178
Johnson A, Herskowitz I (1985) A repressor (*MATα2* product) and its operator control expression of a set of cell type specific genes in yeast. Cell 42:237-247
Nasmyth K (1983) Molecular analysis of a cell lineage. Nature (London) 302:670-676
Nasmyth K (1985) A repetitive sequence that confers cell cycle START (CDC28) dependent transcription of the *HO* gene in yeast. Cell 42:225-235
Osley MA, Gould J, Kim S, Kane M, Hereford L (1986) Identification of sequences in a yeast histone promoter involved in periodic transcription. Cell 45:537-544
Osley MA, Hereford L (1982) Identification of a sequence responsible for periodic synthesis of yeast histone 2A mRNA. Proc Natl Acad Sci USA 79:7689-7693
Peterson T, Prakash L, Prakash S, Osley MA, Reed SI (1985) Regulation of *CDC9*, the *Saccharomyces cerevisiae* gene that encodes DNA ligase. Mol Cell Biol 5:226-235
Plumb M, Stein J, Stein G (1982) Co-ordinate regulation of multiple histone mRNAs during the cell cycle in HeLa cells. Nucleic Acids Res 11:2391-2410
Sittman D, Graves RA, Marzluff NF (1983) Histone mRNA conentrations are regulated at the level of transcription and mRNA degradation. Proc Natl Acad Sci USA 80: 1849-1853
Smith MM, Andresson OS (1983) DNA sequences of yeast H3 and H4 histone genes form two non-allelic gene sets encode identical H3 and H4 proteins. J Mol Biol 169: 663-690
Struhl K (1982) Regulatory sites for *HIS3* gene expression in yeast. Nature (London) 300:284-287

The Determination of Mother Cell-Specific Mating Type Switching in Yeast by a Specific Regulator of *HO* Transcription

K. Nasmyth[1]

Introduction

Mating type switching in homothallic yeast is tightly regulated. It occurs almost every division in certain a or α haploid cells but not at all in diploid a/α cells (Hawthorne 1963, Takano and Oshima 1970, Hicks and Herskowitz 1976). In addition to its repression in diploids, switching is tightly regulated also during the growth of haploid cells. It occurs during a restricted phase of the mother cell cycle, but not at all during the cell cycle of daughter cells (Hicks and Herskowitz 1976, Strathern and Herskowitz 1979). This paper addresses the question of how switching is confined to mother cells.

Switching is thought to be initiated by a specific double-strand cleavage of mating type DNA by an endonuclease encoded by the *HO* gene (Strathern et al. 1982, Kostricken et al. 1983). Two lines of evidence suggest that the failure of daughter cells to switch is due to the absence of *HO* endonuclease in such cells. First, neither the endonuclease activity nor its mRNA are made during the first cell cycle of G_0 stationary phase daughters upon their inoculation into fresh medium, whereas they are both produced in a cell cycle-dependent fashion in mother cells under the same circumstances (Nasmyth 1983). Moreover, neither the endonuclease nor its mRNA are sufficiently stable to be transferred to a daughter cell after their synthesis in a mother cell.

The second piece of evidence concernes the analysis of cells carrying a recombinant *HO* gene whose transcription is driven by the GAL upstream promoter element, which is presumably active in both mother and daughter cells (Jensen 1983, Jensen and Herskowitz 1984). In the presence of galactose, which activates the GAL upstream promoter element (Guarente et al. 1982), daughter cells were now observed to switch as frequently as mother cells. It is, therefore, a reasonable working hypothesis, that the failure of daughter cells to switch their mating type is due to an underlying failure to transcribe the *HO* gene throughout their cell cycle.

There are at least two different types of approach to the question of why mother and daughter cells adopt different patterns of transcription during their cell cycles. The first, which one might describe as a "top down" approach, would involve studying the physiological differences between mother and daughter cells which might be significant to the asymmetry of *HO* transcription. Daughter cells, for instance, are born somewhat smaller than mother cells and consequently have a longer G_1 period (Pringle and Hartwell 1981). New cell wall and membrane material is deposited predominantly in the bud (i.e., the emerging daughter cell). Proteins are predominantly secreted in the bud (Schekman and

[1]Medical Research Council, Laboratory of Molecular Biology, Hills Road, Cambridge CB2 2QH, England

37. Colloquium Mosbach 1986
Cell Cycle and Oncogenes

Norvick 1981), whereas the chitinous scar produced during septation is inherited exclusively by the mother cell. Finally, arteficial circular minichromosomes lacking a centromere are segregated predominantly to the mother cell nucleus at mitosis (Murray and Szostak 1983). A priori, this last difference, involving the behaviour of DNA, would seem to be the most relevant to the asymmetry of *HO* transcription.

Even if it were known whether any of these differences between mother and daughter cells were significant to the asymmetry of *HO* transcription, it would still be necessary to determine how such physiological states were communicated to the factors that directly affect transcription of *HO*. This consideration leads one to a second, "bottom up", approach to the phenomenon. Here, the emphasis is to identify those factors which bind to the *HO* promoter whose presence, alteration, or absence is responsible for the differential activity in mother and daughter cells. The experiments described in this paper concern the putative identification of a gene called *SWI5* which may encode such a factor. Our data suggest that *SWI5* is a sequence-specific DNA binding protein which is required for activation of the *HO* promoter. Though *SWI5* is only one of several gene products required for efficient transcription of *HO*, it may be the only one whose activity is missing in daughter cells.

Results

A GAL/HO Hybrid Promoter

In addition to a TATA like region at -90, the *HO* promoter requires sequences between -1000 and -1400 (which are called *URS1* (Nasmyth 1985a). In contrast, the DNA between -900 and -150 (called *URS2*) can be deleted without loss of promoter function, but it is required for correct cell cycle control (Nasmyth 1985b). Removal of *URS2* leaves an active promoter which is still apparently sensitive to the Mother/Daughter state. This is confirmed by the data in Table 1, which shows that switching caused by such a deletion (Δ229.102) still occurs predominantly in mother cells. It has been suggested, therefore, that *URS1* contains DNA sequences which activate *HO* only in mother cells (Nasmyth 1985a). This hypothesis predicts that replacement of *URS1* by an upstream activation sequence from another promoter which is not sensitive to the mother/daughter state might cause the *HO* promoter to become active in daughter cells.

Such a hybrid promoter has been made here by replacing the 400 bp of *HO* DNA between -1000 and -1400 by a 330 bp DNA fragment containing the GAL *UAS* (see Fig. 1). Two types of such a construction have been made. The hybrid promoter has been linked either to an intact *HO* structural gene or to an *ho*-β-galactosidase gene fusion (see Fig. 1). Both types have been transferred to the *HO* locus using the *SUP4-0* replacement technique described by Nasmyth (1985a).

The *GAL/HO* promoter gives rise to *HO* transcripts when cells are grown on galactose but not when they are grown on glucose. The level of *HO* RNA produced by the hybrid promoter (when induced on galactose) is similar to that produced by a wild-type *HO* promoter in cells grown on either glucose or galactose (see Table 2). This suggests that the *GAL UAS*, which is only active in the presence of galactose, can fulfill the function normally executed by *URS1*, the wild-type *HO* UAS.

Table 1. Pedigree of switching caused by deletion no. 229.102

	Divisions	α→2a switches	α→1a,1α switches
Mothers	96	47	0
Daughters	93	2	0
Spores	35	0	0

The pedigree of switching caused by the 229.102 deletion was determined as described by Strathern and Herskowitz (1979). An a/α diploid homozygous for *HO* Δ229.102 was sporulated and the spores dissected in the vicinity of a heavy streak of α cells on a YEPD plate. *MATa* spores germinate, undergo G1 arrest, and cell enlargement, whereas *MATα* spores undergo one or more cell divisions in the presence of α factor supplied by the α cell streak. Switching is detected when an α cell that has divided in the presence of α factor produces a pair of cells that undergo G1 arrest and cell enlargement.

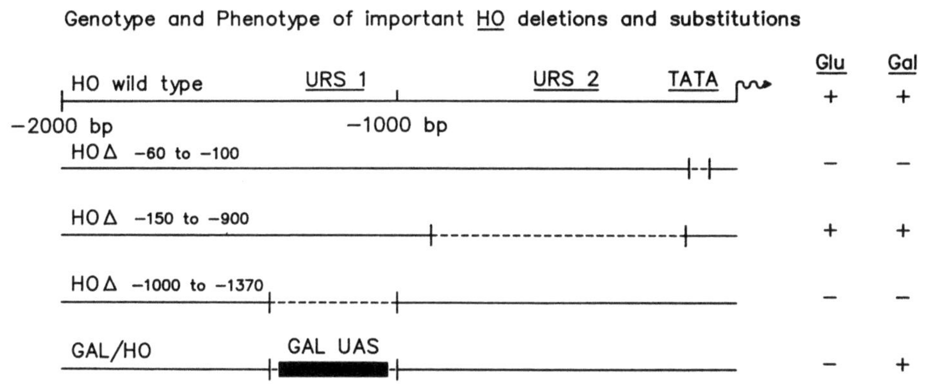

Fig. 1. The genotype and phenotype of *HO* deletions and substitutions. On the left hand side of the diagram, *HO* promoter DNA is represented by a *solid line*, deleted DNA by a *dotted line*, and inserted DNA by a *solid bar*. On the right hand side of the diagram, the levels of *HO* RNA produced in YEP glucose or YEP galactose are represented by + or - signs (+ being >100 x -). All constructions were made and transplaced into the *HO* locus in yeast as described by Nasmyth (1985)

Table 2. The relative levels of *HO* RNA produced by the *HO* and *GAL/HO* promoters in glucose and galactose medium

Promoter	*HO*	*GAL/HO*
Glucose	100	< 3.0
Galactose	82.6	137

HO RNA levels were measured by SI mapping as described by Nasmyth (1985). Cells were grown either in YEP glucose or YEP galactose medium.

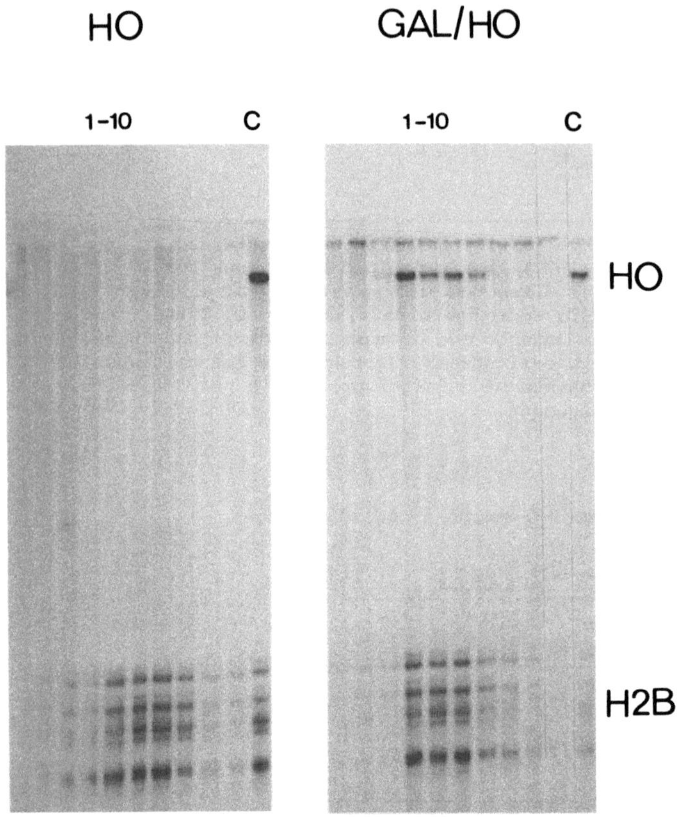

Fig. 2. The level *HO* RNA from *HO* and *GAL/HO* in daughter cells in YEP galactose. The strains bearing *HO (left)* and *GAL/HO (right)* contained the ts *cdc15-2* mutation. Stationary phase daughter cells were prepared from cells grown at the permissive temperature 25°C and innoculated into fresh YEP galactose medium at 37°C and samples were collected after 30, 60, 90, 120, 150, 180, 225, 270, 315, and 360 min (lanes *1-10*). An aliquot of cells was also grown at 25°C for 580 min (lane marked *c*). At 37°C the G1 cells will undergo only a single (daughter) cell cycle and will arrest in a late stage of mitosis, whereas at 25°C the cells will proceed into subsequent cell cycles (i.e., daughter and mother cell cycles). The level of *HO* and H2B RNA's are shown

Regulatory Properties of *GAL/HO*

The galactose dependent *HO* transcription from the hybrid *GAL/HO* promoter is still subject to two of the three physiological controls that affect the wild-type *HO* promoter. *HO* transcription is repressed by the a/α state and is cell cycle-dependent (data not shown; see also Fig. 2). The third major physiological control that affects the wild-type *HO* promoter is its activity in mother cells but not in daughter cells. This property has been previously observed by measuring the level of *HO* transcripts during the first cell cycle of outgrowth of G_0 stationary phase daughter cells (Nasmyth 1983, 1985a). Figure 2 shows that, unlike the wild-type *HO* promoter, the *GAL/HO* hybrid produces abundant *HO* transcripts in a cell cycle-dependent manner in such daughter cells when grown on galactose. A comparison of the *HO* RNA levels found during

Table 3. The number of switches per cell division caused by *HO* and *GAL/HO*

	Mothers	Daughters	Spores
HO	62/96 (64.6%)	0/88 (0)	0/18 (0)
GAL/HO	84/162 (51.9%)	58/152 (38.2%)	18/58 (31%)

The pedigrees were determined as described in the legend to Table 1, except that cells were grown on a plate concerning YEP galactose.

the first cell cycle of isolated daughter cells with those found in mixed population of mother plus daughter cells suggests that the level of transcription in daughter cells is similar to that in mother cells (Fig. 2). These data are consistent with the hypothesis that the *GAL* UAS, which is presumably active in both mother and daughter cells, is substituting for a mother cell-specific *HO* UAS which normally resides in *URS1*.

Mating Type Switching Caused by *GAL/HO*

Three major rules for yeast mating type switching have emerged from the direct observation of switching cells under the microscope (Strathern and Herskowitz 1979): The first, and possibly most important, is the rule that a cells do not switch their mating type when arrested in G1 by α factor. Indeed, it is this property that allows the facile determination of the mating type of individual cells undergoing switching. The second rule is that cells always switch in pairs; that is, both progeny of a switching cell have the new mating type. The third rule is that only cells that have previously budded (i.e., mother cells) can switch their mating type. The question arises, therefore, as to whether *HO* transcription driven by the hybrid *GAL/HO* promoter causes mating type switching that disobeys any of these rules. To answer this, a/α diploids homozygous either for *HO* or *GAL/HO* were sporulated and the spores allowed to germinate on galactose medium. Table 3 summarizes the types of α to a mating type switches observed. In both cases, all switches occurred in pairs and both a cells produced remained arrested by α factor for several hours. As expected, the wild-type *HO* promoter failed to cause any spore or daughter cell to switch. In contrast, the *GAL/HO* promoter caused daughter cells (and spores) to switch their mating type almost as frequently as mother cells. For the *HO* strain, mother cells switched their mating type at 65% of their cell divisions, whereas daughter and spores not at all. For the *GAL/HO* strain, on the other hand, mother cells switched mating type at 52% of their divisions, daughter cells at 38%, and spores at 31%. Examples of the type of pedigrees obtained with the *GAL/HO* strains are shown in Fig. 3.

Since the *GAL/URS* replacement of *URS1* in the *HO* promoter leads to less than a twofold increase in *HO* transcription which is still cell cycle regulated, it is reasonable to conclude from the above results that the acquisition by daughter cells of the sort of *HO* transcription that normally occurs only in mother cells is sufficient to cause them also to switch their mating types efficiently. In other words, the failure of normal daughter cells to switch their mating types must be mainly determined by their lack of *HO* endonuclease, which is due to their failure to transcribe the *HO* genes.

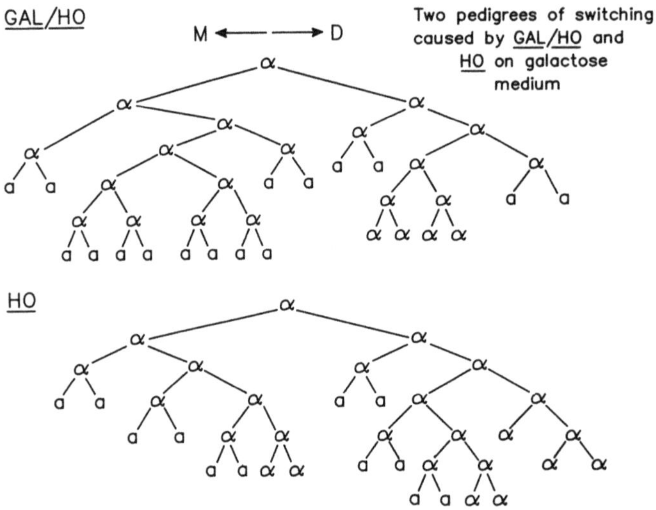

Fig. 3. Actual pedigrees produced by the *HO* and *GAL/HO* promoters. This pedigree analysis was performed as described in the legend of Table 1. Of each pair of sister cells produced at division, the mother cell is shown on the left and the daughter on the right. Cells were inferred to have an α mating type if they divided in the presence of α factor and a's if they underwent cell cycle arrest and massive "Schmoo" like enlargement

Which Regulators are Responsible for Mother/Daughter Control of *HO* Transcription?

The differential transcription of *HO* in mother and daughter cells could be due either to a daughter cell-specific repressor or a mother cell specific activator (or both). Mutation of a putative daughter specific repressor will lead merely to transcription in both mother and daughter cells, a phenotype that is difficult to detect in any mutant screen. Mutation of a putative mother specific activator, on the other hand, should lead to a total defect in *HO* transcription, a phenotype which can be readily detected.

So far, ten unlinked genes called *SWI*, 1, 2, 3, 4, 5, 6, 7, 8, 9, and 10 (Haber and Garvick 1977, Stern et al. 1984, Breeden and Nasmyth, data not shown) have been identified as being necessary for *HO* transcription. Formally speaking, the ten *SWI* genes are all activators of *HO* transcription. The question is whether any of them are active only in mother cells. An analysis of the dependence of the *GAL/HO* promoter upon the various *SWI* genes is revealing in this regard. Table 4 compares the level of β-galactosidase activity due either to the wild-type *HO* promoter (in glucose) or the *GAL/HO* hybrid promoter (in galactose) in wild type and mutants in each of the ten *SWI* genes. It is striking that the *GAL/HO* promoter is still noticeably dependent upon *SWI* 1, 2, 3, 4, 6, 7, 8, 9, and 10 but not upon *SWI* 5. The implication is that all the *SWI* gene products, with the possible exception of *SWI5*, must be to some extent (if not fully) active in daughter cells. *SWI5* is, therefore, the only candidate for a mother cell-specific activator.

The logic behind the above conclusion is as follows. If x% of *GAL/HO* transcription is *SWI* X independent and if *SWI* X is exlusively active in mother cells, then at most x% of transcription can be independent of the mother/daughter state. This value, which can be derived from the proportion of *GAL/HO* transcription in swi^- and SWI^+ strains, varies from 1% for *swi* 4-100 to 100% for *swi* 5-100 (see Table 2). The actual figure for the proportion of *GAL/HO* transcription that is independent of the mother/daughter state is hard to quantitate precisely, but the data of Fig. 2 suggests a conservative estimate of no less than 50%. Thus, no *SWI* gene for which x is significantly less than 50% can be exclusively

Table 4. Levels of ho-β-galactosidase activity from the *HO* and *GAL/HO* promoters in *swi*⁻ and *SWI*⁺ strains

	HO (15-20 ū)	*GAL/HO* (50-60 ū)
SWI⁺	100	100
swi1-2	< 1.0	< 5.0
swi2-314	0.14	2.5
swi3-1	< 1.0	< 5.0
swi4-10	< 0.1	< 1.0
swi5-100	0.69	105.0
swi6-399	< 0.1	2.0
swi7-339	2.6	11.3
swi8-299	2.4	16.0
swi9-289	3.4	15.7
swi10-281	0.58	4.4

The construction of the *swi* mutant strains used here will be described elsewhere. In the case of strains containing the *ho*-βgal fusion, cells were grown in YEP glucose. In the case of strains containing the GAL/*ho*-βgal fusion, cells were grown in YEP galactose. β-galactosidase activities were measured as described by Guarente et al. (1982).

active in mother cells. This includes *SWI* 1, 2, 3, 4, 6, 7, 8, 9, and 10, but not *SWI5*, which is apparently not at all required for *GAL/HO* transcription and is therefore the only serious candidate for a mother cell specific activator.

Analysis of the *SWI5* gene

The previous experiments have singled out the *SWI5* gene product as a possible mother cell specific activator of *HO*. In order to pursue the function of *SWI5*, we have cloned and sequenced the *SWI5* gene (Bankier and Nasmyth, data not shown). An uninterrupted open reading frame suggests that *SWI5* encodes a polypeptide of 709 amino acids, of which a 100 amino acid long region at the C terminus bears striking homology to the *Xenopus* 5S gene transcription factor TfIIIa. TfIIIa has recently been shown to be composed of nine tandem 30 amino acid subdomains, each of which is thought to bind a single atom of zinc (Miller et al. 1985a). Figure 4 shows that the *SWI5* gene possesses three tandem domains which are strikingly homologous to the consensus sequence of the TfIIIa repeat. The highly conserved residues include a tyrosine or phenylalanine at position 6, a pair of cysteines at positions 8 and 13, a phenylalanine at position 17, a leucine or isoleucine at position 23, and a pair of histidines at positions 26 and 30. Miller et al. (1985a) have proposed that a single atom of zinc is tetrahedrally coordinated with the pairs of cysteines and histidines and that the intervening polypeptide chain forms a finger that interacts with the phosphate backbone and major or minor grooves of DNA. The TfIIIa repeat is a highly conserved protein subdomain since it has also been found recently in the Kruppel (Rosenberg et al. 1986) and Serendipity (Vincent et al. 1985) genes of *D. melanogaster* and the *ADR1* gene of yeast (Hartshorne et al. 1986).

TfIIIa possibly has two functions. First, it is a sequence-specific DNA binding protein which is necessary for the assembly of a stable

```
1             6   8             13             17                23          26             30

T G E K * P Y V C . . D G C D K R F T K K . . L K * R H . . * . H     TfIIIa
                                                                      Consensus

M P D K * T F E C L F P G C T K T F K R R Y N I R * S H I Q T * H
                                                                      SWI5
L E D R * P Y S C D H P G C D K A F V R N H D L I * R H K K S * H     residues
                                                                      544 to 632
* Q E K * A Y A C P * * * C G K K F N R E D A L V * V H R S R M I C
```

Fig. 4. A comparison of the predicted amino acid sequence of the C terminal domain of *SWI5* and the consensus sequence of the repeated subdomain of TfIIIa. *Stars* represent spaces and *dots* represent actual non conserved residues. (Data courtesy of A. Bankier)

transcription complex on the 5S gene. Second, it has a less well-defined role in the metabolism of 5S RNA itself. In oocytes, most of the TfIIIa protein is found as a specific 7S complex with 5S RNA itself (Pelham and Brown 1980). The homology of *SWI5* to TfIIIa suggests, therefore, that it may play similar roles in the transcription of *HO* and in the metabolism of its RNA. Preliminary results (D. Stillman, personal communication) suggest that *SWI5* is indeed a sequence specific DNA binding protein with at least one binding site within the *URS* 1 region of the *HO* promoter. We do not yet know, however, whether *SWI5* also plays a role in the metabolism of *HO* RNA.

Discussion

This paper describes the construction and characterization of a hybrid promoter composed by replacing the *HO* upstream activation sequence by that of the *GAL* promoter. The hybrid *GAL/HO* promoter is only active in cells grown on galactose, suggesting that the *GAL UAS* is playing an essential role, but it nevertheless retains the transcriptional start points, the a/α repression, and cell cycle dependence characteristic of the parental *HO* promoter.

A crucial difference between the *HO* and *GAL/HO* promoters is that the latter is active in daughter cells whereas the former is not. *GAL/HO* produces in both mother and daughter cells the sort of *HO* transcription normally (i.e., from *HO*) only seen in mother cells. Daughter cells normally never switch their mating type, but as a consequence of the alteration in *HO* transcription caused by *GAL/HO*, they now switch almost as frequently as mother cells. This implies that transcription of *HO* is the major rate limiting event for switching which is normally absent in daughter cells. It should be pointed out that the *GAL/HO* promoter appears to cause mother cells to switch slightly more frequently than daughter cells: 52% versus 38%. It is not known whether this is a reflection of some residual asymmetry in *GAL/HO* transcription or due to another process involved in switching being less efficient in daughter cells.

The observation that daughter cells will switch their mating type as efficiently as mother cells when *HO* transcription is deregulated has been reported previously by Jensen and Herskowitz (1984). The crucial difference between that experiment and this one is that the fusion used by Jensen and Herskowitz (1984) causes a hundredfold increase in *HO* RNA levels, whereas the fusion described here produces a level of *HO* RNA in daughter cells that is close to that normally seen in mother cells. In addition, the changed switching pattern due to the present *GAL/HO* promoter cannot be attributed to any changed cell cycle control.

Why is *HO* Only Transcribed in Mother Cells?

A cororally of the above conclusions is that *HO* is normally never transcribed during the G1 period of log phase daughter cells. How is this asymmetry in *HO* transcription between mother and daughter cell achieved? One possibility is that an activator essential for *HO* transcription is present only in mother cells. The isolation of mutants which are defective in *HO* transcription has identified ten so-called *SWI* genes, any one of which could in principle encode a mother cell-specific activator. The data presented in this paper appear to exclude, however, nine out of the ten *SWI* genes from this role.

The only gene for which there is no evidence that it has a role in daughter cells is *SWI5*. The *GAL/HO* promoter does not require *SWI5* at all. It is, therefore, a tenable working hypothesis that the asymmetry of *HO* transcription is due to asymmetric distribution of *SWI5* activity between mother and daughter cells. Analysis of the DNA sequences of *SWI5* suggests that it encodes a protein which contains, at its C terminus, three regions homologous to the repeated DNA binding domain of the 5S transcription factor TfIIIa. It is likely, therefore, that *SWI5* activates *HO* by binding directly to some part of its promoter.

Implications for Combinatorial Control

The regulatory behaviour of the *GAL/HO* fusion has significance for the general question of combinatorial control. The activation of promoters only under a special combination of conditions is clearly a central feature of the differentiation of eukaryotic cells. The *HO* promoter in yeast is a good example: it is only active in certain cell types (a or α but not a/α), only when the cell undergoes Start, and only if the cell is a mother cell. Distinct sequences have been implicated in a/α repression (Miller et al. 1985b) and cell cycle regulation (Nasmyth 1985b). The observation that the *GAL/HO* promoter is insensitive to the mother/daughter state but retains cell type and cell cycle dependence implies that a third set of different DNA sequences are involved in mother/daughter control.

We do not know how the three controls are integrated on the *HO* promoter. It seems that the system must be modular because sensitivity to the mother/daughter state can be replaced by sensitivity to the presence of galactose by replacing the *HO* UAS (called *URS1*) by an equivalent region of the GAL promoter, without prejudicing the other two *HO* controls. A modular nature to complex combinatorial controls would of course greatly facilitate their evolution.

One of the striking facts concerning hybrid *GAL/HO* promoters is that the *GAL* UAS does not require *SWI2* or *SWI4* when placed 150 bp upstream of the *HO* RNA 5' terminus (Stern et al. 1984), yet it acquires a strong dependence when placed 1000 bp upstream. Continuing this theme, it should be stressed that the *GAL* UAS, though normally only 200 to 300 bp upstream of the start sites of *GAL1* and *GAL10*, is still capable of action when placed 1200 bp upstream from the *HO* start site. Presumably, this long range *GAL4* effect (*GAL4* is the protein responsible for *GAL UAS* activity) is only possible due to facilitation by *SWI* gene products acting via sequences within *URS2*. Elucidating the molecular mechanism behind such phenomena may shed considerable light upon the nature of combinatorial control.

One of the current models for cooperative long-range regulation proposes that stable transcription complexes can be formed through the interaction of different factors via protein-protein contacts while each protein is still bound to specific sites which may be hundreds or even thousands of base pairs apart. Recent observations of the cooperative binding of lac (Besse et al. 1986) or Lambda repressor (Hochschild and Ptashne 1986) to distant pairs of operators have confirmed that such a model is feasible. It is possible to imagine, therefore, that different *SWI* gene products, e.g., *SWI4* and *SWI5*, manage to interact directly even though they act via different cis acting DNA sequences at *HO*. In this context, it is worth pointing out, however, that *GAL4* can also cooperate with *SWI4* even though there has almost certainly been no previous history of such an interaction. This implies that any interaction between *SWI5* and *SWI4* is unlikely to be via highly specific protein-protein contacts and may therefore involve a third party such as an RNA polymerase subunit.

References

Besse M, Von Wilcken-Bergmann B, Müller-Hill B (1986) Synthetic lac operator mediates repression through lac repressor when introduced upstream and downstream from the lac promoter. EMBO Journal 5:1377-1382

Guarente L, Rogers Yocum R, Gifford P (1982) A GAL10-CYC1 hybrid yeast promoter identifies the *GAL4* regulatory region as an upstream site. Proc Natl Acad Sci USA 79:7411-7414

Haber JE, Garvick B (1977) A new gene affecting the efficiency of mating type interconversions in homothallic strains of Saccharomyces cerevisiae. Genetics 87:33-50

Hartshorne TA, Blumberg H, Young ET (1986) Sequence homology of the yeast regulatory protein *ADR1* with Xenopus transcription factor TfIIIa. Nature (London)

Hawthorne DC (19863) Directed mutation of the mating type alleles as an explanation of homothallism in yeast. Abstr Proc 11th int Congr Genet 7:34-35

Hicks JB, Herskowitz I (1976) Interconversion of yeast mating types. I. Direct observation of the action of the homothallism (HO) gene. Genetics 83:245-258

Hochschild A, Ptashne M (9186) Cooperative binding of Lambda repressors to sites separated by integral turns of the DNA helix. Cell 44:681-687

Jensen R (1983) Control of mating type inter-conversion in the yeast Saccharomyces cerevisiae. Ph.D. Univ Oregon

Jensen RE, Herskowitz I (1984) Directionality and Regulation of Cassette substitution in yeast. Cold Spring Harbor Symp Quant Biol XL1X 97-104

Kostriken R, Strathern JN, Klar AJS, Hicks JB, Heffron F (1983) A site-specific endonuclease essential for mating-type switching in Saccharomyces cerevisiae. Cell 35:167-174

Miller J, McLachlan AD, Klug A (1985a) Repetitive zinc-binding domains in the protein transcription factor IIIA from Xenopus oocytes. EMBO Journal 4:1609-1614

Miller AM, Mackay VL, Nasmyth KA (1985b) Identification and comparison of two sequence elements that confer cell-type specific transcription in yeast. Nature (London) 314:598-603

Murray AW, Szostak JW (1983) Pedigree analysis of plasmid segregation in yeast. Cell 34:961-970

Nasmyth KA (1983) Molecular analysis of a cell lineage. Nature (London) 302:670-676

Nasmyth K (1985a) At least 1400 base pairs of 5' flanking DNA is required for the correct expression of the *HO* gene. Cell 42:213-223

Nasmyth K (1985b) A repetitive sequence that confers cell cycle START *(CDC28)* dependent transcription of the *HO* gene in yeast. Cell 42:225-235

Pelham HR, Brown DD (1980) A specific transcription factor that can bind either the 5S RNA gene or 5S RNA. Porc Natl Acad Sci USA 77:4170-4174

Pringle JR, Hartwell LH (1981) The Saccharomyces cerevisiae cell cycle. In: The Molecular Biology of the yeast Saccharomyces. Edit. Strathern et al. Cold Spring Harbor Laboratory

Rosenberg et al. (1986) Structural homology of the product of the Drosophila Krüppel gene with Xenopus transcription factor IIIa. Nature (London) 319:336-339

Schekman R, Norvick P (1981) The secretory process and yeast cell surface assembly. Strathern et al. (eds) Cold Spring Harbor Laboratory

Stern M, Jensen R, Herskowitz I (1984) Five SWI genes are required for expression of the HO gene in yeast. J Mol Biol 178:853-868

Strathern JN, Hershowitz I (1979) Asymmetry and directionality in production of new cell types during clonal growth: the switching pattern of homothallic yeast. Cell 17:371-381

Strathern JN, Klar AJS, Hicks JB, Abraham JA, Ivy JM, Nasmyth KA, McGill C (1982)
Homothallic switching of yeast mating-type cassettes is initiated by a double-stranded cut in the MAT locus. Cell 31:183-192

Takano I, Oshima Y (1979) Mutational nature of an allele specific conversion of the mating type by the homothallism gene HOa in Saccharomyces. Genetics 65:421-427

Vincent A, Colot HV, Rosbash M (1985) J Mol Biol 186:149-166

II. Cellular Oncogenes and Their Relation to the Cell Cycle

Analysis of the *neu*-Encoded Protein and Its Mechanism of Oncogenic Activation

C. I. Bargmann[1], D. F. Stern[1], J. Drebin[2], A. Schechter[1], and R. A. Weinberg[1]

The search for oncogenes and their encoded proteins has involved a variety of experimental procedures and reagents. In our own laboratory, the procedure of gene transfer, also known as transfection, has been used extensively in an attempt to deter oncogenes residing in a variety of tumor DNA's. These DNA's have been applied via the Graham-van der Eb calcium phosphate procedure to monolayers of NIH3T3 cells with the hope that some of the transfected cells would respond to donor transforming sequences by demonstrating an altered growth pattern in monolayer. Indeed in this fashion, we and others were able to demonstrate the presence of oncogenes in the DNA's of a variety of tumor cell types (Varmus 1984).

When the results first appeared in the years 1979 to 1981, we had no insight into the identities of the oncogenes themselves or their encoded proteins. Consequently, we resorted to an experiment that we hoped would reveal the nature of the oncogene-encoded proteins. NIH3T3 cells that had been transformed by a variety of transfected oncogenes were expanded to mass cultures and inoculated into semi-syngeneic NFS mice. These cell lines, which arose originally from foci in transfected monolayers, were also tumorigenic and formed rapidly growing fibrosarcomas in the mice. We hoped that the tumor cells, which had acquired a novel, exogenous gene by transfection, would display a novel antigen encoded by the introduced gene and that this antigen would be immunogenic in these NFS mice. Thus we analyzed the sera of the tumor-bearing mice with the hope that they would immunoprecipitate proteins present in the tumor cells that are absent in untransfected NIH3T3 cells (Padhy et al. 1982).

This strategy failed in the great majority of instances. The transfectants bearing a variety of tumor oncogenes, while strongly tumorigenic, almost always failed to elicit a specific tumoral immune response. In retrospect, we realize that most of these tumor oncogenes were members of the *ras* gene family. These *ras* genes encode proteins of 21,000 MW that are highly conserved evolutionarily and are poorly immunogenic. An exception to this came from work with a set of oncogenes derived from neuro-glioblastomas induced by others in BDIX by transplacental exposure to the alkylating agent ethyl nitrosourea (ENU).

The ENU used by these workers, when applied to a rat fetus late in development or in the early post-natal period induced these neurectodermal tumors which appeared 3 - 6 months later. We found that NIH3T3 cells transformed by the oncogenes of these tumors invariably displayed at 185,000 MW antigen which was absent in untransfected NIH3T3 cells and was potently immunogenic in the tumor-bearing NFS mice (Padhy et al.

---

[1]Whitehead Institute for Biomedical Research and M.I.T. Department of Biology, Nine Cambridge Center, Cambridge, MA 02142, USA

[2]Department of Pathology, Harvard Medical School

37. Colloquium Mosbach 1986
Cell Cycle and Oncogenes

1982). Such behavior was quite distinct from that of the transfected *ras* oncogenes and suggested that this neuroblastoma oncogene, which we termed *neu*, represented a novel genetic sequence.

We developed an extensive biochemical characterization of the *neu*-induced p185 antigen (Stern et al. 1986). Antibodies against this protein were able to stain intact cells, indicating that it was a cell surface protein bearing extracellular determinants. The protein was found to be glycosylated and phosphorylated. However, in spite of these studies, we had no idea about the role of this protein in cellular physiology nor about its connection with the transfected *neu* oncogene. Experimental strategies to clone the gene seemed difficult and very labor-intensive.

A breakthrough came with the discovery by Downward, Waterfield and their colleagues of a connection between another cell surface protein, the epidermal growth factor receptor (EGFr) and a well-studied retrovirus oncogene, borne by avian erythroblastosis virus and termed *erbB*. Sequencing of the EGFr showed strong homology with the protein encoded by *erbB* and thereby demonstrated the principal that a growth factor receptor protein, or at least a version of it, can serve as an agent of cellular transformation (Downward et al. 1984). Indeed, as further work would demonstrate, the *erbB* oncogene represents a truncated version of the chicken EGFr gene which lacks both its right and left ends (Ullrich et al. 1984, Liberman et al. 1985).

Such findings suggested an explanation of our own results, in that the EGFr had at least superficial similarity with the oncogene-induced p185. As our own work rapidly confirmed, an *erbB* oncogene probe reacted strongly with a novel DNA fragment present in all NIH3T3 cells that had acquired the *neu* oncogene by transfection (Schechter et al. 1985). This suggested that *neu* was a version of the EGFr (i.e., *erbB*) gene and that p185 was an altered version of the EGFr protein. Such a notion was supported by use of polyclonal antisera reactive with the human EGFr protein. Such antisera also demonstrated some reactivity with p185 displayed on *neu*-transfected NIH3T3 cells (Schechter et al. 1985).

The relationship of the two antigens was studied by more direct biochemical analysis. Here it was found that the two proteins differed: upon gel electrophoresis, the EGFr migrated as a protein having a mass 15,000 less than that of p185 (Stern et al. 1986). This left the possibility that the two proteins were related but distinct, being specified by different members of a gene family. Alternatively, it was possible that p185 was an altered form of the EGFr, having acquired its distinct properties as a consequence of the genetic changes that led to the activation of the *neu* oncogene.

One resolution of this problem came from analysis of the proteins displayed by rat-1 cells, these being immortalized rat fibroblasts which appeared to carry no transforming oncogene. These rat-1 cells displayed both p185 and the EGFr on their surface, indicating that p185 could be synthesized in the apparent absence of any oncogenic activating mutation (Stern et al. 1986). It appeared to be a normal cell surface constituent of the cell as was the EGFr also displayed by these cells. This observation also yielded a second, related conclusion – that the p185 displayed by *neu*-transfected NIH3T3 cells had a counterpart of virtually identical structure in normal rat cells. This suggested in turn that the events that led to activation of the *neu* gene had little effect on the structure of the protein (p185) that seemed to be encoded by this gene.

The observed homology of *erbB* and *neu* made it possible for ourselves and others to isolate genomic and cDNA versions of this gene. In our own laboratory, 33 kb cosmid clones of the genomic versions *neu* oncogene and of the normal rat *neu* protooncogene were isolated (Hung et al. 1986). The oncogene clone was found to be biologically active in that it could induce foci in NIH3T3 monolayers. Importantly, the oncogenic and normal *neu* cosmid clones appeared identical in their restriction enzyme cleavage patterns. This further supported the notion that the activating lesions caused only minor changes in the structure of the protooncogene.

Such results were compatible with two possibilities. Either the oncogene became activated by lesions affecting its level of expression or subtly affecting the structure of the encoded protein. One of us developed an NIH3T3 line that strongly overexpressed the normal *neu* cosmid clone. These over-expressing cells were morphologically normal, leading us to conclude that deregulation of expression was not the mechanism through which the *neu* oncogene achieved its transforming powers (Hung et al. 1986). Later work, described below, supported the alternative conclusion that a lesion subtly affecting protein structure was responsible for the observed transforming properties of the gene.

While much of this was underway, a parallel line of work arose from experiments that led to the isolation of monoclonal antibodies against the p185 protein. This work, of Jeffrey Drebin and Mark Greene, then at the Harvard Medical School, yielded several antibodies that reacted only with the rat but not mouse versions of this protein. Such work supported the notion that the immunogenicity of p185 in mice stemmed in large part from the fact that it was a rat protein encoded by an introduced rat gene (Drebin et al. 1984).

We decided to use these monoclonal antibodies to alter the behavior of *neu*-transformed cells. Unlike most other oncogenes, which encode proteins that are localized to the cytoplasm or nucleus, the *neu* and *erbB* oncogenes specify cell surface proteins. Because of this, we reasoned that antibodies against these oncogene-encoded proteins might affect the growth properties of the cells transformed by these oncogenes.

Such expectations were realized. The anti-p185 monoclonal antibody was able to completely inhibit the anchorage-independent growth of *neu*-transfected NIH3T3 (Drebin et al. 1985). *ras*-transformants, by contrast, were unaffected by the antibody. This growth inhibition was reversible, in that removal of antibody from the culture medium led to outgrowth of agar colonies which soon reached the size of untreated colonies. This and other results indicated that the effects of the antibody were cytostatic rather than cytotoxic. In consonance with this was an analysis of the antibody molecules which indicated that they were poor in fixing complement.

The antibody-mediated inhibition of cell growth appeared to depend upon the cross-linking of surface p185. Monovalent Fab antibody fragments, made by limited proteolysis of the antibody molecules, retained the ability to stain specifically the transfected cells in an immunofluorescence test, but these fragments were unable to inhibit cell growth in agar. When these Fab fragments were relinked by means of antimouse immunoglobulin, they then reacquired the ability to inhibit anchorage-independent growths (Drebin et al. 1985).

The growth-inhibitory effects of the monoclonal antibody were paralleled by a strong reduction in the amount of p185 detected in the antibody-treated cells. We suspect that the effects of the antibody derive from its ability to crosslink surface p185. This in turn induces cell surface clustering of the antigen molecules followed by their internalization and degradation. In this way, the transfected cell is deprived of the protein that has been driving its growth. In addition this close association of p185 down-modulation with loss of anchorage-independent growth underscores the central role played by this protein in mediating the oncogene-induced transformation phenotypes.

Further study of p185 depended upon a more detailed understanding of the primary sequence of this protein. This information came from a cDNA clone that was made from *neu*-specific RNA in a transfected cell (Bargmann et al. 1986a). The cDNA appeared to encode the entirety of the p185 antigen. Its expression could be driven by fusion with the strong constitutive promoter derived from the SV40 genome. Cells transfected with this chimeric gene expressed full length p185 and became strongly transformed.

The sequence analysis of the cDNA clone showed that it encodes a protein that has many structural similarities to the EGFr. Both proteins have extracellular domains with identical arrays of cysteines, although the other amino acid residues in this domain are divergent (Bargmann et al. 1986). Both proteins have hydrophobic segments that appear to function as transmembrane domains and threonine residues in the cytoplasmic domains that could serve as substrates for protein kinase C. The proteins have very similar cytoplasmic domains that encode tyrosine kinase functions. Such analysis vindicated earlier suspicions that the two proteins are encoded by divergent members of a single gene family.

The structural analysis also makes it clear that p185 functions as a growth factor receptor. The nature of its ligand remains obscure, however. We have tested a variety of the well-characterized growth factors for their ability to induce changes in p185, either down-modulation or autophosphorylation. None has shown a specific affinity for p185. Cells that highly overexpress the normal rat p185 demonstrate no increased affinity for binding EGF, showing that p185 must respond to a different, distinct ligand. We are currently trying a variety of tests in an attempt to identify the ligand, which may represent a novel growth factor (Stern et al. 1986).

The sequence analysis of the oncogenic *neu* cDNA made it possible to explore the mechanism of activation. The evidence presented earlier suggested that the activating lesion affected the structure of the encoded protein, not its level of expression. As such, we reasoned that the lesion should be represented in the cDNA sequence and could be uncovered by comparative analysis of the oncogenic and normal cDNAs. Accordingly, a normal *neu* cDNA was isolated. Although apparently identical in overall structure to the transforming cDNA, it lacked transforming powers when introduced into an expression vector. This confirmed our suspicions that the activating lesion lay within the cDNA sequence and was a subtle one.

We then undertook a program of interallelic recombination between the two cDNA's, attempting in this way to localize the activating mutation to a particular portion of the sequence. Such recombination experiments showed that the activating lesion was to be found in a region of the gene that encodes a segment of p185 surrounding and including the transmembrane domain.

It remained only to sequence the normal cDNA in this region. The answer was quick to come and unambiguous (Bargmann et al. 1986a). A valine residue present in the transmembrane domain of the normal protein was replaced by a glutamic acid in the oncogene-encoded protein. This was most unexpected in that the transmembrane domain of growth factor receptors has been suspected of serving only a rather passive role in anchoring the protein in the plasma membrane. We suspect that the activating lesion causes the protein to be constitutively active, emiting stimulatory signals even in the absence of bound ligand. The normal protein, we speculate, is able to send out such signals only after encountering the appropriate ligand. This reasoning in turn implicates the transmembrane domain as playing a central role in the signal transduction mechanism utilized by the normal protein in passing information from outside the cell into the cytoplasm. It remains unclear how such a simple amino acid change, which introduces a hydrophilic amino acid into the normally hydrophobic transmembrane domain, is able to alter so profoundly the physiology of a 185,000 MW protein.

The *neu* oncogene in which this point mutation was found was only one of four *neu* oncogenes that had been detected in the DNA of chemically induced rat neuroblastomas. Three other tumor cell lines, also developed by workers at the Salk Institute at La Jolla, had been shown five years earlier to contain the *neu* oncogene (Shih et al. 1981). The nature of the activating mutations that created these other independently arising oncogenes was also of interest.

We speculated that these other lesions might also affect the same region of the *neu* protooncogene as was found altered in our initial analysis. Consequently, we utilized radiolabeled oligonucleotides, the sequences of which were related to this portion of the transmembrane-encoding sequence in order to probe the DNA's of a series of cell lines which carried these other *neu* oncogenes. These oligonucleotide probes were used under hybridization conditions which would allow them to anneal to genomic DNA carrying *neu* oncogenes having a sequence perfectly complementary to that of the probes; conversely, such probes would not anneal if they encountered even a one nucleotide mismatch.

The outcome of these experiments was the finding that these three independently arising oncogenes carry the identical activating mutation to that found in the initially studied gene. In all cases, a T to A transversion was observed. Such results are perhaps unexpected for two reasons. First, such transversion mutations are not predicted to occur frequently after ethylnitrosourea-induced alkylation. Instead, transitions seem to be found. Second, it is surprising that a randomly acting mutagen like ENU, which must react with many millions of nucleotides in the genome, repeatedly induces tumors carrying the identical activating mutation. Clearly, strong selective pressures must intervene to ensure that only those cells bearing a very limited number of activating lesions are favored to grow out into tumors. One can only speculate about the nature of these selective pressures at present.

The *neu* oncogene has thus, quite unexpectedly, provided insight into a number of unrelated problems, yielding information in the areas of growth factor receptors, chemical carcinogenesis, and tumor cell immunology. Ultimately, we hope that a detailed study will reveal the identity of its ligand, a hitherto elusive growth factor, and shed light on the metabolic mechanisms by which the oncogene-encoded protein is able to transform cells.

References

Bargmann CI, Hung M-C, Weinberg RA (1986a) Multiple independent activations of the *neu* oncogene by a point mutation altering the transmembrane domain of p185. Cell (submitted)

Bargmann CI, Hung M-C, Weinberg RA (1986b) The *neu* oncogene encodes an EGF receptor-related protein that may be a growth factor receptor. Nature (London) 319:226-230

Downward J, Yarden Y, Mayes E, Scrace G, Totty N, Stockwell P, Ullrich A, Schlessinger J, Waterfield MD (1984) Close similarity of epidermal growth factor receptor and v-*erb*-B oncogene protein sequences. Nature (London) 307:521-527

Drebin JA, Stern DF, Link VL, Weinberg RA, Greene MI (1984) Monoclonal antibodies identify a cell-surface antigen associated with an activated cellular oncogene. Nature (London) 312:545-548

Drebin JA, Link VC, Stern DF, Weinberg RA, Greene MI (1985) Down-modulation of an oncogene protein product and reversion of the transformed phenotype by monoclonal antibodies. Cell 41:695-706

Hung M-C, Schechter AL, Chevray P-YM, Stern DF, Weinberg RA (1986) Molecular cloning of the *neu* gene: absence of gross structural alteration in oncogenic alleles. Proc Natl Acad Sci USA 83:261-264

Liberman TA, Nusbaum HR, Razon N, Kris R, Lax I, Soreq H, Whittle N, Waterfield MD, Ullrich A, Schlessinger J (1985) Amplification, enhanced expression and possible rearrangement of EGF receptor gene in primary human brain tumors of glial origin. Nature (London) 313:144-147

Padhy LC, Shih C, Cowing D, Finkelstein R, Weinberg RA (1982) Identification of a phosphoprotein specifically induced by the transforming DNA of rat neuroblastomas. Cell 28:865-871

Schechter AL, Hung M-C, Vaidyanathan L, Weinberg RA, Yang-Feng T, Francke U, Ullrich A, Coussens L (1985) The *neu* gene: an *erbB*-homologous gene distinct from and unlinked to the gene encoding the EGF receptor. Science 229:976-978

Shih C, Padhy LC, Murray M, Weinberg RA (1981) Transforming genes of carcinomas and neuroblastomas introduced into mouse fibroblasts. Nature (London) 290:261-264

Stern DF, Heffernan PA, Weinberg RA (1986) p185, product of the *neu* proto-oncogene, is a receptor-like protein associated with tyrosine kinase activity. Mol Cell Biol (in press)

Ullrich A, Coussens L, Hayflick JS, Dull TJ, Gray A, Tam AW, Lee J, Yarden Y, Libermann TA, Schlessinger J, Downward J, Mayes ELV, Whittle N, Waterfield MD, Seeburg PH (1984) Human epidermal growth factor receptor cDNA sequence and aberrant expression of the amplified gene in A431 epidermoid carcinoma cells. Nature (London) 309:418-425

Varmus HE (1984) The molecular genetics of cellular oncogenes. Annu Rev Genet 18: 553-612

c-*fos* and Growth Control

R. Müller and R. Bravo[1]

Introduction

It is now generally accepted that the normal cellular homologues of retroviral oncogenes, the proto-oncogenes, may have crucial physiological functions in the regulation of cellular proliferation. This hypothesis is strongly supported by the observation that three proto-oncogenes, c-*sis*, c-*erb*B and c-*fms* are part of the cellular growth factor-receptor system (Doolittle et al. 1983, Waterfield et al. 1985, Downward et al. 1984, Sherr et al. 1985). In addition, expression of proto-oncogenes encoding nuclear proteins has been shown to increase rapidly following the stimulation of different cell types using several mitogenic agents: The transient transcriptional activation of the c-*fos* gene is followed by the accumulation of c-*myc* RNA (Kelly et al. 1983, Greenberg and Ziff 1984, Cochran et al. 1984, Kruijer et al. 1984, Müller et al. 1984). This article summarizes the available data suggesting an early role for c-*fos* in inducing the "competent state" in fibroblasts and thus in the movement of quiescent cells into the cell cycle. In addition, we will discuss the signal transduction pathways that seem to be involved in the induction of c-*fos* by growth factors.

Induction of c-*fos* mRNA and Protein by Growth Factors

Stimulation of quiescent (i.e., serum-deprived) NIH3T3 cells with fetal calf serum (FCS) leads to the synchronous entry of the cell population into the S-phase of the cell cycle within 10 - 12 h (Müller et al. 1984). The earliest known change in gene expression following stimulation of fibroblasts with FCS is the induction of c-*fos* transcription (Cochran et al. 1984, Greenberg and Ziff 1984, Kruijer et al. 1984, Müller et al. 1984). c-*fos* mRNA levels increase at least 50-fold within 30 min and decrease rapidly to basal levels within the following 90 min (Fig. 1). Purified growth factors, such as fibroblast growth factor (FGF) and platelet-derived growth factor (PDGF) are equally as efficient at inducing c-*fos* expression as is 10% FCS (Müller et al. 1984). Although the induction of c-*fos* precedes the accumulation of c-*myc* mRNA, it is unlikely that the c-*fos* gene product is involved in the induction of c-*myc*, as both genes can be induced in the presence of the protein synthesis inhibitor cycloheximide (Cochran et al. 1984, Greenberg and Ziff 1984, Müller et al. 1984).

The induction of c-*fos* mRNA is rapidly followed by the synthesis of c-*fos* protein. Maximum synthesis is observed approximately 1 h after stimulation, and after another hour c-*fos* protein synthesis is greatly reduced, correlating with the observed levels of c-*fos* mRNA. The growth

[1]European Molecular Biology Laboratory, Postfach 10.2209, D-6900 Heidelberg, FRG

37. Colloquium Mosbach 1986
Cell Cycle and Oncogenes

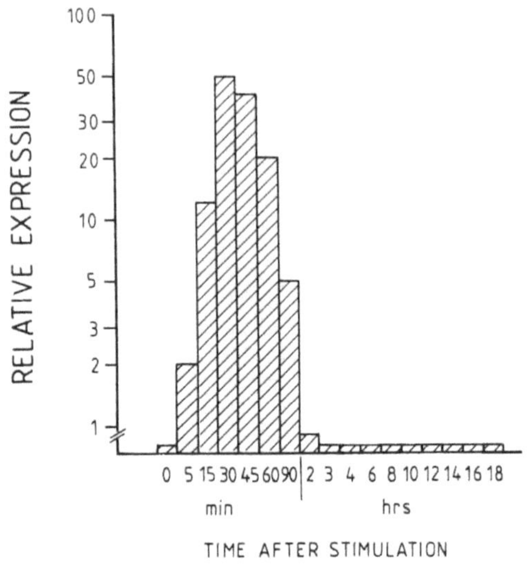

Fig. 1. Relative expression of c-*fos* mRNA following the stimulation of quiescent NIH3T3 cells with FCS. Quantification of data shown in Müller et al. (1984)

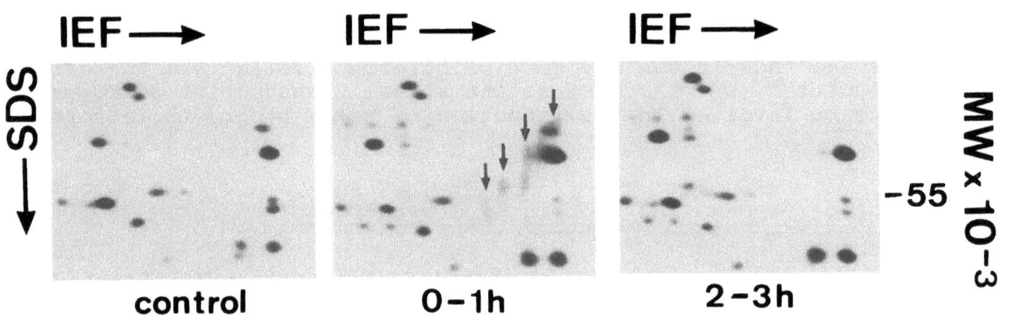

Fig. 2. Two-dimensional gel analysis of c-*fos* protein in FCS-stimulated NIH3T3 cells. Quiescent cells were stimulated with 10% FCS and labeled for 1 h with [^{35}S]-methionine at the indicated times. Only the area of interest of the two-dimensional gel is shown. *Arrows* indicate the position of *fos* protein

factor-induced c-*fos* protein is extensively post-transcriptionally modified, as judged from its increased molecular mass. While unmodified c-*fos* protein migrates as a 54 kDa protein on reducing polyacrylamide gels, several forms in the range of 55 - 65 kDa are observed in stimulated 3T3 cells (Müller et al. 1984, Kriujer et al. 1984). As judged from two-dimensional gel analysis of c-*fos* protein, the increase in molecular weight is associated with a drop in the isoelectric point (Fig. 2). The post-translational modifications are much less extensive in transforming v-*fos* gene products and may therefore represent a regulatory mechanism controlling the oncogenic potential of c-*fos* protein (Müller et al. 1984). The nature of these modifications, except for phosphorylation of serine and threonine residues, remains obscure.

c-*fos* Expression Is Very Low During the Cell Cycle But May Play a Role in Competence Induction

Recent analyses have shown that c-*fos* mRNA levels are extemely low or even undetectable throughout the cell cycle in NIH3T3 cells (Bravo et al. 1986). This suggests that a high expression of c-*fos* is not required for the continuous cycling of NIH3T3 cells, and may therefore be restricted to special conditions. A role of c-*fos* in the normal proliferation of cells, however, cannot be completely ruled out, since c-*fos* expression in growing cells has been shown to be slightly elevated compared to quiescent cells (Müller et al. 1984).

It has been shown that growth factors can act as either "competence factors" or "progression factors" (Stiles et al. 1979). Quiescent mouse fibroblasts briefy exposed to FGF or PDGF become competent to synthesize DNA. The addition of platelet-poor plasma (PPP) which contains progression factors but lacks PDGF allows these cells (but not untreated cells) to progress through G1 and enter the S-phase (Stiles et al. 1979). There are several lines of evidence that support the hypothesis that the c-*fos* gene product may play a role in conferring competence on stimulated cells. First, both c-*fos* and c-*myc* are induced specifically by competence factors, not by the progression factors in PPP (Bravo et al. 1985). Second, it has been shown that the induction of competence is possible at different stages during the cell cycle (Stiles et al. 1979). Accordingly, c-*fos* and c-*myc* can be induced by PDGF or serum in any phase of the cell cycle except for mitosis (Bravo et al. 1986). Third, wounding of a confluent monolayer of cells induces competence in cells lining the wound, as shown by their mitogenic response to progression factors (Stiles et al. 1979). Likewise, cells lining the wound transiently express c-*fos* protein (Verrier et al. 1986). On the other hand, cells transformed by c-*fos* are not able to grow in PPP (R. Bravo and R. Müller, unpublished observations) suggesting that c-*fos* and c-*myc* may function in concert to promote cell division. This conclusion is supported by the observation that cells transfected with c-*myc* show a competence-like state, but the growth response to progression factors is weak relative to PDGF (Armelin et al. 1984). The fact that several other genes are induced by growth factors or serum in quiescent cells (Cochran et al. 1983, Lau and Nathans 1985) suggests that the transition from Go to G1 is the result of a complex interaction of several genes including c-*fos* and c-*myc*.

The competence state induced by a brief PDGF treatment is very stable. Competent cells show a significant decrease in their response to progression factors only 8 hours after the withdrawal of the competent factor. However, PDGF-stimulated cells kept in serum-free medium for a similar time express c-*fos* at a level similar to that observed in quiescent cells (Bravo et al. 1985). When PPP is added to these cells, only a weak induction of c-*fos*, but a strong mitogenic effect is observed. The persistence of the competent state cannot be due to the presence of high levels of c-*fos* protein since its half-line is approximately 30 min (Müller et al. 1984), contrasting the half life of 16 h for competence (Singh et al. 1982, Bravo et al. 1985). These observations suggest the persistence of competence could be due to the presence of a stable biochemical change triggered by the early induced genes.

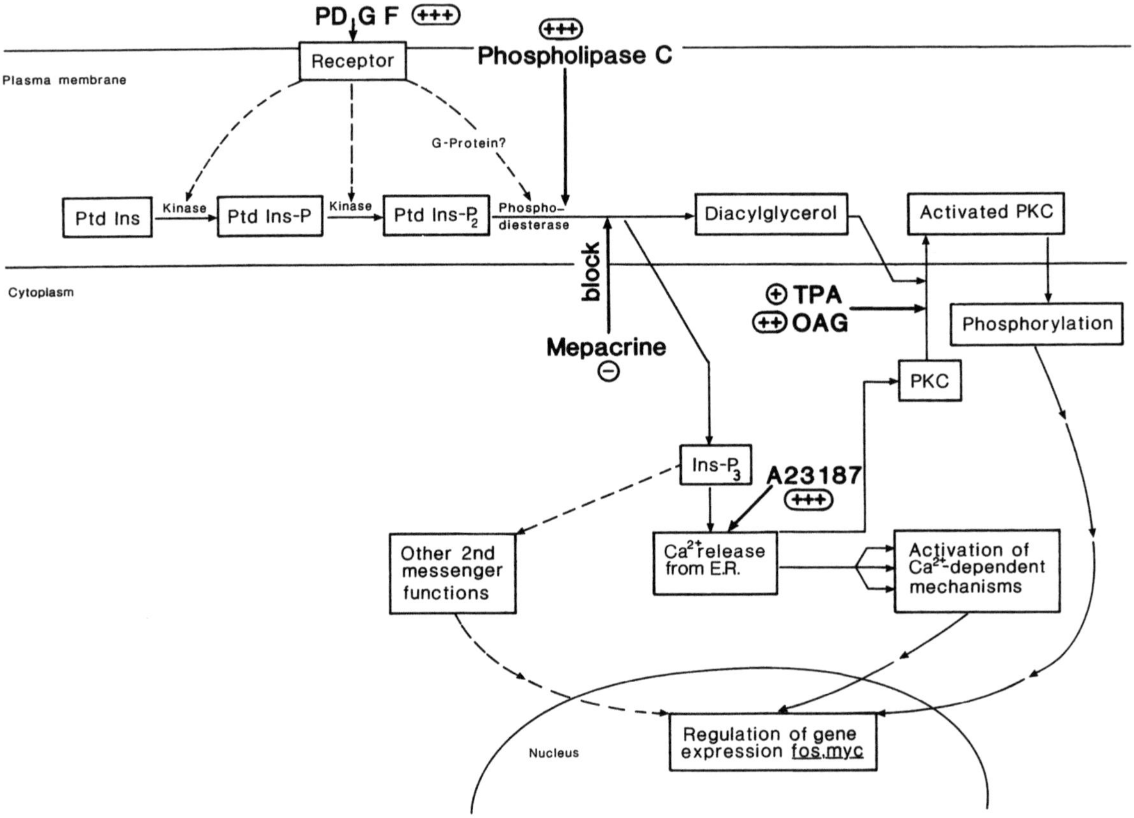

Fig. 3. Schematic representation of early events following the stimulation of fibroblasts with a growth factor *(GF)* such as PDGF. The relative induction of c-*fos* by various compounds affecting phospholipid degradation, PKC activation and Ca^{2+} release is shown in bold script using a + scale. (+++), induction observed by 10 ng of PDGF/ml; (++), 1/3 to 1/5 of +++; (+), approximately 1/10 of +++; (-), more than tenfold inhibition of the PDGF induced expression of c-*fos*

Involvement of Phospholipid Degradation, Protein Kinase C and Ca^{2+} in c-*fos* Induction

To elucidate the mechanism of c-*fos* induction, it is of obvious importance to understand the mechanisms and pathways transducing signals from occupied growth factor receptors to the c-*fos* gene. Binding of a growth factor to its receptor leads to the immediate phosphorylation of the receptor and the activation of its intrinsic kinase activity. This is rapidly followed by pleiotropic events, including the degradation of phospholipids, phosphorylation of specific cellular proteins, and changes in protein transport, intracellular pH, and cyclic nucleotide metabolism. A central role in transduction of growth factor signals has been suggested for phosphoinositol and the breakdown products of its phosphorylated forms (for reviews see Nishizuka 1984, Berridge and Irvine 1984). Upon binding of a growth factor, phosphoinositol is rapidly converted in two steps to phosphoinositol 4,5-diphosphate (PtdIns-P2), which is subsequently cleaved to yield diacylglycerol and inositol 1,4,5-triphosphate (Ins-P3) (see Fig. 1). While diacylglycerol leads to the activation of protein kinase C (PKC), Ins-P3 triggers the release of Ca^{2+} from the endoplasmic reticulum (for reviews see Nishizuka 1984, Berridge and Irvine 1984) (see also Fig. 3). To elucidate the pathways involved in the transduction of growth factor signals to the c-*fos* gene we used several compounds known to affect phospholipid degradation, activation of PKC and release of intracellular Ca^{2+}. We also made use of the observation that the constant exposure (for several days) of 3T3 cells to the tumor promoter 12-O-tetradecanoylphorbol-13-acetate (TPA), a phorbol ester able to stimulate PKC due to its structural homology with diacylglycerol, leads to the down-regulation of PKC (Nishizuka 1984). The results of these studies showed that the degradation of PtdIns-P2 and the subsequent activation of PKC and release of Ca^{2+} seem to play an important role in c-*fos* induction (see Fig. 3). This conclusion is based on several observations. (1) c-*fos* is induced by phospholipase C, an enzyme that specifically cleaves PtdIns-P2 to yield diacylglycerol and Ins-P3. In agreement with this finding, c-*fos* induction by PDGF is abolished by mepacrine, an inhibitor of the enzymatic degradation of phospholipids. (2) The tumor promoter TPA, as previously described, and the diacylglycerol analogue OAG induce c-*fos*, albeit to lowr levels than PDGF. These observations suggest a role for PKC but also point to the existence of other mechanisms (see point 4 below), involved in signal transduction to the c-*fos* gene. (3) A similar conclusion is reached from the finding that c-*fos* induction by PDGF is decreased, but not abolished in NIH3T3 cells pretreated for 3 days with TPA and thus depleted of PKC. (4) There seem to be other pathway(s) involved in c-*fos* induction which are controlled by intracellular Ca^{2+} levels. When NIH3T3 cells were treated with high concentrations (10 µM) of the calcium ionophore A23187 c-*fos* induction was almost as efficient as with PDGF or FCS. This strong induction is presumably due to the fact that at high Ca^{2+} concentrations PKC is activated even in the absence of diacylglycerol. However, induction of c-*fos* to lower levels was also observed in TPA-pretreated cells, suggesting a role for Ca^{2+} in signal-transducing mechanisms other than the activation of PKC.

Another point that deserves attention if the fact that OAG is a more efficient inducer of c-*fos* in both fibroblasts and macrophages than TPA. Although other reasons cannot be ruled out, it is possible that PKC is not the only target for OAG while TPA acts specifically on PKC. In this context it is interesting to note that several genes have been isolated which are distinct from, but closely related to the gene encoding PKC.

References

Armelin HA, Armelin MCS, Kelly K, Stewart T, Leder P, Cochran BH, Stiles CD (1984) Functional role for c-*myc* in mitogenic response to platelet-derived growth factor. Nature (London) 310:655-660

Berridge MJ, Irvine RF (1984) Inositol triphosphate, a novel second messenger in cellular signal transduction. Nature (London) 312:315-321

Bravo R, Burckhardt J, Müller R (1985) Persistence of the competent state in mouse fibroblasts is independent of c-*fos* and c-*myc* expression. Exp Cell Res 160:540-543

Bravo R, Burckhardt J, Müller R (1986) Expression of c-*fos* in NIH3T3 cells is very low but inducible throughout the cell cycle. EMBO J 5:695-700

Cochran BH, Reffel AC, Stiles CD (1983) Molecular cloning of gene sequences regulated by platelet-derived growth factor. Cell 33:939-947

Cochran BH, Zullo J, Verma IM, Stiles CD (1984) Expression of the c-*fos* oncogene and a newly discovered r-*fos* is stimulated by platelet-derived growth factor. Science 226:1080-1082

Doolittle RF, Hunkapillar MW, Hood LE, Deware SG, Robbins KC, Aaronson SA, Antoniades HN (1983) Simian sarcoma virus *onc* gene, v-*sis*, is derived from the gene (or genes) encoding a platelet-derived growth factor. Science 221:275-277

Downward J, Yarden Y, Mayes E, Scrace G, Totty N, Stockwell P, Ullrich A, Schlessinger J, Waterfield MD (1984) Close similarity of epidermal growth factor and v-*erB* oncogene protein sequence. Nature (London) 307:521-527

Greenberg ME, Ziff EM (1984) Stimulation of 3T3 cells induces transcription of the c-*fos* proto-oncogenes. Nature (London) 311:433-438

Kelly K, Cochran BH, Stiles CD, Leder P (1983) Cell-specific regulation of the c-*myc* gene by lymphocyte mitogens and platelet-derived growth factor. Cell 35:603-610

Kruijer W, Cooper JA, Hunter T, Verma IM (1984) Platelet-derived growth factor induces rapid but transient expression of the c-*fos* gene and protein. Nature (London) 312:711-716

Lau LF, Nathans D (1985) Identification of a set of genes during the Go/G1 transition of cultured mouse cells. EMBO J 4:3145-3151

Müller R, Bravo R, Burckhardt J, Curran T (1984) Induction of c-*fos* gene and protein by growth factors precedes activation of c-*myc*. Nature (London) 312:716-720

Nishizuka Y (1984) The role of protein kinase C in cell surface signal transduction and tumour promotion. Nature (London) 308:693-698

Sherr CJ, Rettenmier CW, Sacca R, Roussel MF, Look AT, Stanley ER (1985) The c-*fms* proto-oncogene product is related to the receptor for the mononuclear phagocyte growth factor, CSF-1. Cell 41:665-676

Singh JP, Chaikin MA, Pledger WJ, Scher CD, Stiles CD (1982) Persistence of the mitogenic response to platelet-derived growth factor (competence) does not reflect a long-term interaction between the growth factor and the target cell. J Cell Biol 96:1497-1502

Stiles CD, Capone GT, Scher CD, Antoniades HN, Van Wyk JJ, Pledger WJ (1979) Dual control of cell growth by somatomedins and platelet-derived growth factor. Proc Natl Acad Sci USA 76:1279-1283

Verrier B, Müller D, Bravo R, Müller R (1986) Wounding a fibroblast monolayer results in the rapid induction of the c-*fos* proto-oncogene. EMBO J 5:913-917

Waterfield MD, Scrace GT, Whittle N, Stroobant P, Johnsson A, Westeson A, Westermark B, Helding C-H, Huang JS, Deuel TF (1983) Platelet-derived growth factor is structurally related to the putative transforming protein $p28^{sis}$ of simian sarcoma virus. Nature (London) 304:35

Structure and Function of *myc* and *mil* Oncogenes

K. Bister[1]

Introduction

The transformation-specific oncogenes (or v-*onc* genes) of highly oncogenic retroviruses represent transduced mutant alleles of cellular oncogenes whose normal nonmutated alleles are termed protooncogenes (or c-*onc* genes). Protooncogenes are obviously nononcogenic in their normal cellular environment, and most of them presumably fulfil essential physiological functions in cell growth and development. The transition from a normal chromosomal allele with physiological function to a transduced mutant allele with oncogenic function is always accompanied by changes in gene structure and control of gene expression. However, it is largely unknown which of the qualitative or quantitative changes in gene structure and expression are necessary or even sufficient for oncogenic activation. Furthermore, there is as yet no unequivocal identification of the relevant biochemical function(s) of the protein products of both the mutant (transduced) and the normal (chromosomal) allele of an oncogene. Hence, the molecular mechanisms of cell transformation by v-*onc* genes are still obscure.

This chapter will focus specifically on recent findings on the biochemistry and molecular genetics of two particular oncogenes, termed *myc* and *mil*, which were originally identified as transforming genes of highly oncogenic avian retroviruses. More comprehensive reviews on these and other oncogenes have been published recently (Varmus 1984, Bister 1986, Bister and Jansen 1986).

The *myc* Oncogene

Specific protein products and nucleic acid sequences of the *myc* oncogene were originally discovered for the transduced allele in the genome of acute leukemia virus MC29 (Bister et al. 1977, Duesberg et al. 1977). The discovery of *myc* alleles in the genomes of three other independently isolated leukemia viruses (CMII, OK10, MH2) with pathogenic properties similar to those of MC29 provided strong genetic evidence that *myc* represents the common oncogenic determinant of these viruses (Bister and Duesberg 1980). The search for the origin of *myc* led to the identification of the cellular c-*myc* gene which proved to be highly conserved among vertebrate species and which is now suspected to be possibly implicated in viral or nonviral animal and human malignancies (Varmus 1984, Bister 1986, Bister and Jansen 1986).

The c-*myc* gene is composed of three exons and two intervening sequences. The open reading frame encoding a 416-amino-acid protein ex-

[1]Otto-Warburg-Laboratorium, Max-Planck-Institut für Molekulare Genetik, D-1000 Berlin 33 (Dahlem), FRG

37. Colloquium Mosbach 1986
Cell Cycle and Oncogenes

tends from a translational initiation codon within the second exon to a termination codon within the third exon. The corresponding protein product identified in extracts from normal chicken cells has an apparent molecular weight of 58,000. The four naturally transduced mutant alleles of the chicken c-*myc* gene were found in different genetic contexts in the genomes of the four highly oncogenic independent retroviral isolates MC29, CMII, OK10, and MH2. All four v-*myc* alleles contain nontruncated and precisely fused coding regions from the second and third exon of the cellular gene, but lack any sequences from the first (noncoding) exon and various 3' complements from the untranslated region of the third exon. The MC29, OK10, and MH2 v-*myc* alleles also contain sequences derived from the first c-*myc* intron. Within the coding domains, the transduced alleles differ from their common cellular progenitor by various nucleotide substitutions and, in the case of MH2 v-*myc*, by a small in-frame deletion (see below). The MC29, CMII, and OK10 v-*myc* alleles are fused in frame with virion genes and are expressed via genome-sized mRNAs as *gag-myc* (MC29, CMII) or *gag-pol-myc* (OK10) hybrid proteins with apparent molecular weights of 110,000, 90,000, or 200,000, respectively. The OK10 v-*myc* allele is also expressed via subgenomic mRNA as a 60,000-molecular-weight protein which presumably contains 11 amino-terminal amino acid residues specified by the first six *gag* codons present in the leader sequence and five codons derived from the noncoding 5' terminus of the second c-*myc* exon, in addition to 416 amino acid residues encoded by sequences derived from the c-*myc* coding region. The MH2 v-*myc* allele is expressed via subgenomic mRNA as a 59,000/61,000-molecular-weight protein whose amino terminus is presumably structured like that of the OK10 60,000-molecular-weight protein. MH2 carries another cell-derived oncogene, v-*mil*, which is expressed via genome-sized mRNA as a *gag-mil* hybrid protein (see below). In summary, all natural oncogenic retroviral isolates containing transduced *myc* alleles carry and express genetic information other than authentic c-*myc* coding sequences; specifically, the protein products of all transduced *myc* alleles differ from the c-*myc* protein by their modified amino termini and by various substitutions and (for MH2) a small deletion of amino acid residues (Bister and Jansen 1986). Hence, in the absence of a classical genetic definition, a rigorous biochemical definition of sequences necessary and/or sufficient for cell transformation by v-*myc* oncogenes could not be directly deduced from the genetic structure of these viruses. Based on molecular and biological analyses of the v-*myc* alleles of CMII and of spontaneous

<u>Fig. 1.</u> Predicted amino acid sequences of protein products of c-*myc* and v-*myc* alleles. The complete sequence of the 416-amino-acid protein product of the chicken (ck) c-*myc* gene is shown. For v-*myc* protein products, amino acid sequences are shown only at positions where they differ from the c-*myc* protein sequence by substitutions or deletions of amino acid residues. Substituted amino acid residues are shown in parentheses when the corresponding mutation of the particular v-*myc* allele was reported in one, but not in other publications. Also, when independent sequence analyses have reported different mutations of the same codon, the corresponding substituted amino acid residues are shown separated by a slash. Note, that amino-terminal sequences of v-*myc* protein products specified by virion genes or by sequences derived from the first c-*myc* intron (see text) are not shown here. The comparison is based on the following published sequences: chicken c-*myc* (Watson et al. 1983), MH2 v-*myc* (Kan et al. 1984, Jansen et al. 1985, Patschinsky et al. 1986a), MC29 v-*myc* (Alitalo et al. 1983, Reddy et al. 1983), OK10 v-*myc* (Hayflick et al. 1985), CMII v-*myc* (Walther et al. 1985, 1986). (A, alanine; R, arginine, N, asparagine; D, aspartic acid; C, cysteine; Q, glutamine; E, glutamic acid; G, glycine; H, histidine; I, isoleucine; L, leucine; K, lysine; M, methionine; F, phenylalanine; P, proline; S, serine; T, threonine; W, tryptophan; Y, tyrosine; V, valine)

										10										20
c-myc(ck)	M	P	L	S	A	S	L	P	S	K	N	Y	D	Y	D	Y	D	S	V	Q
v-myc(MH2)					V															
										30										40
c-myc(ck)	P	Y	F	Y	F	E	E	E	E	E	N	F	Y	L	A	A	Q	Q	R	G
v-myc(MH2)														(S)						S
										50										60
c-myc(ck)	S	E	L	Q	P	P	A	P	S	E	D	I	W	K	K	F	E	L	L	P
										70										80
c-myc(ck)	T	P	P	L	S	P	S	R	R	S	S	L	A	A	A	S	C	F	P	S
v-myc(MH2)	A							C			N									
v-myc(MC29)	M																			
v-myc(OK10)	A																			
										90										100
c-myc(ck)	T	A	D	Q	L	E	M	V	T	E	L	L	G	G	D	M	V	N	Q	S
										110										120
c-myc(ck)	F	I	C	D	P	D	D	E	S	F	V	K	S	I	I	I	Q	D	C	M
v-myc(MH2)	S																R			
										130										140
c-myc(ck)	W	S	G	F	S	A	A	A	K	L	E	K	V	V	S	E	K	L	A	T
										150										160
c-myc(ck)	Y	Q	A	S	R	R	E	G	G	P	A	A	A	S	R	P	G	P	P	P
v-myc(MH2)		K																		
v-myc(MC29)						(Q)														
										170										180
c-myc(ck)	S	G	P	P	P	P	P	A	G	P	A	A	S	A	G	L	Y	L	H	D
										190										200
c-myc(ck)	L	G	A	A	A	A	D	C	I	D	P	S	V	V	F	P	Y	P	L	S
v-myc(MH2)							G			G	S						C			G
										210										220
c-myc(ck)	E	R	A	P	R	A	A	P	P	G	A	N	P	A	A	L	L	G	V	D
v-myc(MH2)	R		G	-	-	-	-					G								
										230										240
c-myc(ck)	T	P	P	T	T	S	S	D	S	E	E	E	Q	E	E	D	E	E	I	D
v-myc(MH2)	A				V/A	G	G	G												
										250										260
c-myc(ck)	V	V	T	L	A	E	A	N	E	S	E	S	S	T	E	S	S	T	E	A
v-myc(MH2)																	(G)			
										270										280
c-myc(ck)	S	E	E	H	C	K	P	H	H	S	P	L	V	L	K	R	C	H	V	N
v-myc(MH2)															E					
										290										300
c-myc(ck)	I	H	Q	H	N	Y	A	A	P	P	S	T	K	V	E	Y	P	A	A	K
										310										320
c-myc(ck)	R	L	K	L	D	S	G	R	V	L	K	Q	I	S	N	N	R	K	C	S
v-myc(MH2)													V							
										330										340
c-myc(ck)	S	P	R	T	S	D	S	E	E	N	D	K	R	R	T	H	N	V	L	E
v-myc(MH2)									V											
v-myc(MC29)					L															
v-myc(OK10)															M					
										350										360
c-myc(ck)	R	Q	R	R	N	E	L	K	L	S	F	F	A	L	R	D	Q	I	P	E
v-myc(MC29)										R										
										370										380
c-myc(ck)	V	A	N	N	E	K	A	P	K	V	V	I	L	K	K	A	T	E	Y	V
v-myc(MH2)															R					
										390										400
c-myc(ck)	L	S	I	Q	S	D	E	H	R	L	I	A	E	K	E	Q	L	R	R	R
v-myc(MC29)			L						(K)											
										410										
c-myc(ck)	R	E	Q	L	K	H	K	L	E	Q	L	R	N	S	R	A				
v-myc(MC29)							N													
v-myc(CMII)																E				

and in vitro-constructed derivatives of MH2, we have now obtained evidence (see below) that neither major structural changes, such as in-frame fusion with virion genes or internal deletions, nor specific, if any, missense mutations of the c-*myc* coding region are necessary for oncogenic activation of transduced *myc* alleles.

The entire nucleotide sequence of the transduced v-*myc* allele of molecularly cloned and biologically active proviral DNA of CMII was determined (Walther et al. 1985, 1986). Within the shared coding domains, CMII v-*myc* and chicken c-*myc* differ by a single nucleotide substitution causing an exchange of the carboxyl-terminal amino acid residue (glutamic acid/alanine) in the predicted sequences of the corresponding protein products. The availability of the CMII v-*myc* sequence now allows a complete and detailed comparison between all naturally transduced v-*myc* alleles and their common progenitor, the chicken c-*myc* gene. In Fig. 1, all mutations in predicted amino acid sequences of protein products of v-*myc* alleles in comparison to the c-*myc* protein sequence are listed, based on previous sequence analyses of MC29, OK10, and MH2 v-*myc* alleles and on the newly determined CMII v-*myc* sequence. All transduced *myc* alleles carry missense mutations: one (CMII v-*myc*), two (OK10 v-*myc*), seven or five (MC29 v-*myc*), and 27 or 25 (MH2 v-*myc*, which also suffered an in-frame deletion of four codons) (Fig. 1). Notably, there is no common specific mutation found in all four v-*myc* alleles. Based on the observation that codon 61 of the c-*myc* sequence is changed in the v-*myc* alleles of MC29, OK10, and MH2 (Fig. 1), it has been argued that mutation of that particular codon may be responsible for oncogenic activation (Papas and Lautenberger 1985). However, the analysis of the CMII v-*myc* sequence reveals that mutation of codon 61 is not necessary for activation of the transforming function of transduced *myc* alleles (Walther et al. 1986).

A small retroviral vector carrying an oncogenic *myc* allele was isolated as a spontaneous variant (MH2E21) of avian oncovirus MH2 (Patschinsky et al. 1986a). The MH2E21 genome, measuring only 2.3 kilobases, can be replicated like larger retroviral genomes and hence contains all *cis*-acting sequence elements essential for encapsidation and reverse transcription of retroviral RNA or for integration and transcription of proviral DNA. The MH2E21 genome contains 5' and 3' noncoding retroviral vector elements and a coding region comprised of the first six codons of the viral *gag* gene and of 417 v-*myc* codons. The *gag-myc* junction corresponds precisely to the presumed splice junction on subgenomic MH2 v-*myc* mRNA, the possible origin of MH2E21. Among the v-*myc* codons, the first five are derived from the noncoding 5' terminus of the second c-*myc* exon, and 412 correspond to the c-*myc* coding region. The predicted sequence of the MH2E21 protein product differs from that of the chicken c-*myc* protein by additional 11 amino-terminal residues and by 25 amino acid substitutions and a deletion of four residues within the shared domains (Fig. 1). To investigate the functional significance of these structural changes, the MH2E21 genome was modified in vitro (Patschinsky et al. 1986a). Using oligonucleotide-directed mutagenesis, the *gag* translational initiation codon was inactivated. Furthermore, all but two of the missense mutations were reverted and the deleted sequences were restored by replacing most of the MH2E21 v-*myc* allele by the corresponding segment of the CMII v-*myc* allele which is isogenic to c-*myc* in that region (Fig. 1). The remaining two 5' mutations have not been found in the v-*myc* alleles of avian oncoviruses MC29, CMII, and OK10. Like MH2 and MH2E21, modified MH2E21 (MH2E21m1c1) transforms avian embryo cells. Like c-*myc*, it encodes a 416-amino-acid protein initiated at the *myc* translational initiation codon.

Based on the results described above, we conclude that neither major structural changes, such as in-frame fusion with virion genes or internal deletions, nor specific, if any, missense mutations of the c-*myc* coding region are necessary for activation of the basic oncogenic function of transduced *myc* alleles. The major structural difference between the v-*myc* alleles and the normal chicken c-*myc* allele is the complete (5' and 3') substitution of retroviral for cellular transcriptional control and processing elements in the transduced alleles. Hence, quantitative changes in gene expression rather than qualitative changes in the structure of the gene product may be essential for the activation of the basic oncogenic function of the *myc* oncogene. It is possible, however, and indeed suggested by the observation of reduced transforming efficiency of constructed MH2 derivatives in which most of the extensively mutated MH2 v-*myc* allele (cf. Fig. 1) has been replaced by CMII v-*myc* sequences (Patschinsky et al. 1986a), that mutations in the coding region may be beneficial for enhanced transforming function and possibly play an important role for the evolution of fully tumorigenic v-*myc* oncogenes.

The *mil* Oncogene

Transduced alleles of the *mil(raf)* oncogene were independently discovered in the genomes of avian acute leukemia virus MH2 and murine sarcoma virus (MSV) 3611 (Jansen et al. 1983a,b, Kan et al. 1983, Coll et al. 1983, Rapp et al. 1983). The avian oncogene, termed v-*mil*, and the murine oncogene, termed v-*raf*, were then recognized to be closely related and to represent transduced alleles of cognate cellular genes, the c-*mil(raf)* genes of chicken and mice (Jansen et al. 1984a, Sutrave et al. 1984). The totally independent transductions of cognate cellular genes in the genomes of highly oncogenic retroviruses from different species are a strong indication for the oncogenic potential of a limited set of conserved eukaryotic genes.

In the MH2 genome, the v-*mil* allele is inserted between partial *gag* sequences at its 5' end and intron-derived v-*myc* sequences at its 3' end. Comparison of v-*mil* with the chicken c-*mil* gene showed that in the course of the transduction coding domains from 11 exons of the cellular gene were precisely fused (Jansen and Bister 1985). The 5' terminus of v-*mil* is derived from sequences within the coding domain of an exon, and the 3' terminus corresponds to a position in the untranslated sequence downstream of the translational termination codon shared between v-*mil* and c-*mil*. The v-*mil* allele is expressed as a *gag-mil* hybrid protein with an apparent molecular weight of 100,000. The nucleotide sequence analyses of c-*mil* genomic DNA and cDNA clones have revealed that the cellular gene contains additional 5' coding domains that have not been transduced into the genome of MH2 (Jansen and Bister 1985, C. Trachmann, M. Koenen, and K. Bister, unpublished data). Concordantly, the normal c-*mil* protein product was identified as a 71,000/73,000-molecular-weight protein (Patschinsky et al. 1986b), whereas the v-*mil*-encoded domain of the MH2 *gag-mil* hybrid protein corresponds to a molecular weight of only 42,000 (Sutrave et al. 1984). Hence, in contrast to the complete transduction of the c-*myc* coding region in the v-*myc* alleles, v-*mil* contains only a partial (3') complement of the chicken c-*mil* coding region. In such a case, qualitative changes of the gene product in both structure and function, but possibly also quantitative changes in gene expression, may be responsible for the transition of this oncogene from the physiological to the malignant form.

The nucleotide sequence analysis of the v-*mil* allele has revealed that the predicted amino acid sequence of its protein product shares significant homology with those of gene products from members of the *src* family of oncogenes (Sutrave et al. 1984). While the products of most of these oncogenes are associated with in vitro tyrosine-specific protein kinase activities, phosphorylation of *mil* protein products both in vivo and in vitro involves serine and threonine residues (Ramsay et al. 1982, Sutrave et al. 1984). The functional significance of these phosphorylations is unknown.

Analyses of spontaneous and in vitro-constructed derivatives of the v-*mil* and v-*myc* containing retrovirus MH2 have revealed that v-*myc* represents the basic oncogenic determinant of this virus and that v-*mil* appears to enhance its oncogenic potential (Jansen et al. 1985, Patschinsky et al. 1986a, Graf et al. 1986). In particular, it was shown that v-*mil*$^-$ v-*myc*$^+$ derivatives transform avian fibroblasts with similar efficiency as wild-type MH2, albeit leading to morphological changes and proliferation rates distinct from those of fibroblasts transformed by the wild-type virus (Jansen et al. 1985, Patschinsky et al. 1986a). Furthermore, v-*mil*$^+$ v-*myc*$^+$ viruses are more oncogenic in vivo and induce factor independent growth of transformed macrophages in tissue culture, whereas v-*mil*$^-$ v-*myc*$^+$ derivatives are less oncogenic and macrophages transformed by these mutants are growth factor dependent, and v-*mil*$^+$ v-*myc*$^-$ mutants do not induce apparent transformation of either macrophages or fibroblasts (Graf et al. 1986). Hence, the cotransduction of two independent cellular genes (see below) in a single retroviral vector apparently is of functional significance for the evolution of retroviruses with an exceptionally high oncogenic potential (Bister 1986).

The v-*mil* and v-*myc* alleles of MH2 are derived from unlinked and unrelated cellular genes on the chicken genome (Jansen et al. 1983b, 1984b). Furthermore, nucleotide sequence analysis of the v-*mil*/v-*myc* junction on the MH2 genome has revealed no obvious clues, such as limited sequence homologies between the c-*mil* and c-*myc* alleles at the sites involved in the recombination, to possible mechanisms of the juxtaposition of these two genes (Jansen and Bister 1985). Chromosomal translocations in the tumor cells from which MH2 was isolated may have preceded a possible simultaneous cotransduction of *mil* and *myc*, although independent sequential recombinations between the transducing virus and the two cellular loci are also conceivable.

References

Alitalo K, Bishop JM, Smith DH, Chen EY, Colby WW, Levinson AD (1983) Nucleotide sequence of the v-*myc* oncogene of avian retrovirus MC29. Proc Natl Acad Sci USA 80:100-104

Bister K (1986) Multiple cell-derived sequences in single retroviral genomes. Adv Viral Oncol 6:45-70

Bister K, Duesberg PH (1980) Genetic structure of avian acute leukemia viruses. Cold Spring Harbor Symp Quant Biol 44:801-822

Bister K, Jansen HW (1986) Oncogenes in retroviruses and cells: biochemistry and molecular genetics. Adv Cancer Res 47:99-188

Bister K, Hayman MJ, Vogt PK (1977) Defectiveness of avian myelocytomatosis virus MC29: isolation of long-term nonproducer cultures and analysis of virus-specific polypeptide synthesis. Virology 82:431-448

Coll J, Righi M, Taisne C de, Dissous C, Gegonne A, Stehelin D (1983) Molecular cloning of the avian acute transforming retrovirus MH2 reveals a novel cell-derived sequence (v-*mil*) in addition to the *myc* oncogene. EMBO J 2:2189-2194

Duesberg PH, Bister K, Vogt PK (1977) The RNA of avian acute leukemia virus MC29. Proc Natl Acad Sci USA 74:4320-4324
Graf T, Weizsaecker F von, Grieser S, Coll J, Stehelin D, Patschinsky T, Bister K, Bechade C, Calothy G, Leutz A (1986) V-*mil* induces autocrine growth and enhanced tumorigenicity in v-*myc* transformed avian macrophages. Cell 45:357-364
Hayflick J, Seeburg PH, Ohlsson R, Pfeifer-Ohlsson S, Watson D, Papas T, Duesberg PH (1985) Nucleotide sequence of two overlapping *myc*-related genes in avian carcinoma virus OK10 and their relation to the *myc* genes of other viruses and the cell. Proc Natl Acad Sci USA 82:2718-2722
Jansen HW, Bister K (1985) Nucleotide sequence analysis of the chicken gene c-*mil*, the progenitor of the retroviral oncogene v-*mil*. Virology 143:359-367
Jansen HW, Patschinsky T, Bister K (1983a) Avian oncovirus MH2: molecular cloning of proviral DNA and structural analysis of viral RNA and protein. J Virol 48:61-73
Jansen HW, Rückert B, Lurz R, Bister K (1983b) Two unrelated cell-derived sequences in the genome of avian leukemia and carcinoma inducing retrovirus MH2. EMBO J 2: 1969-1975
Jansen HW, Lurz R, Bister K, Bonner TI, Mark GE, Rapp UR (1984a) Homologous cell-derived oncogenes in avian carcinoma virus MH2 and murine sarcoma virus 3611. Nature (London) 307:281-284
Jansen HW, Trachmann C, Bister K (1984b) Structural relationship between the chicken protooncogene c-*mil* and the retroviral oncogene v-*mil*. Virology 137:217-224
Jansen HW, Patschinsky T, Walther N, Lurz R, Bister K (1985) Molecular and biological properties of MH2D12, a spontaneous *mil* deletion mutant of avian oncovirus MH2. Virology 142:248-262
Kan NC, Flordellis CS, Garon CF, Duesberg PH, Papas TS (1983) Aviàn carcinoma virus MH2 contains a transformation-specific sequence, *mht*, and shares the *myc* sequence with MC29, CMII, and OK10 viruses. Proc Natl Acad Sci USA 80:6566-6570
Kan NC, Flordellis CS, Mark GE, Duesberg PH, Papas TS (1984) Nucleotide sequence of avian carcinoma virus MH2: two potential *onc* genes, one related to avian virus MC29 and the other related to murine sarcoma virus 3611. Proc Natl Acad Sci USA 81:3000-3004
Papas TS, Lautenberger JA (1985) Sequence curiosity in v-*myc* oncogene. Nature (London) 318:237
Patschinsky T, Jansen HW, Blöcker H, Frank R, Bister K (1986a) Structure and transforming function of transduced mutant alleles of the chicken c-*myc* gene. J Virol 59:341-353
Patschinsky T, Schroeer B, Bister K (1986b) Protein product of proto-oncogene c-*mil*. Mol Cell Biol 6:739-744
Ramsay G, Hayman MJ, Bister K (1982) Phosphorylation of specific sites in the *gag-myc* polyproteins encoded by MC29 type viruses correlates with their transforming ability. EMBO J 1:1111-1116
Rapp UR, Goldsborough MD, Mark GE, Bonner TI, Groffen J, Reynolds FH Jr, Stephenson JR (1983) Structure and biological activity of v-*raf*, a unique oncogene transduced by a retrovirus. Proc Natl Acad Sci USA 80:4218-4222
Reddy EP, Reynolds RK, Watson DK, Schultz RA, Lautenberger J, Papas TS (1983) Nucleotide sequence analysis of the proviral genome of avian myelocytomatosis virus (MC29). Proc Natl Acad Sci USA 80:2500-2504
Sutrave P, Bonner TI, Rapp UR, Jansen HW, Patschinsky T, Bister K (1984) Nucleotide sequence of avian retroviral oncogene v-*mil*: homologue of murine retroviral oncogene v-*raf*. Nature (London) 309:85-88
Varmus HE (1984) The molecular genetics of cellular oncogenes. Annu Rev Genet 18: 553-612
Walther N, Lurz R, Patschinsky T, Jansen HW, Bister K (1985) Molecular cloning of proviral DNA and structural analysis of the transduced *myc* oncogene of avian oncovirus CMII. J Virol 54:576-585
Walther N, Jansen HW, Trachman C, Bister K (1986) Nucleotide sequence of the CMII v-*myc* allele. Virology 154:219-223
Watson DK, Reddy EP, Duesberg PH, Papas TS (1983) Nucleotide sequence analysis of the chicken c-*myc* gene reveals homologous and unique coding regions by comparison with the transforming gene of avian myelocytomatosis virus MC29, *dgag-myc*. Proc Natl Acad Sci USA 80:2146-2150

The T Cell Receptor α Chain and Immunoglobulin Genes in Specific Chromosomal Translocations in Human Leukaemias

T. H. Rabbitts, R. Baer and L. Mengle-Gaw[1]

Introduction

Leukaemias and lymphomas are characterized by the presence of chromosomal abnormalities which are often highly specific to a tumour type. Well-documented examples of this phenomenon are the Philadelphia (Ph') chromosome in chronic myeloid leukaemia and the t8;14 (q24;q32) in Burkitt's lymphoma. In the latter case, molecular cloning studies have shown that the c-*myc* proto-oncogene (present at chromosome band 8q24) and the immunoglobulin heavy (H) chain genes (present at chromosome band 14q32) are juxtaposed as a result of the chromosomal rearrangement (Rabbitts 1985). This association results in the activation of the c-*myc* gene and is thought to be, at least partly, the cause of Burkitt's lymphoma (BL). In addition to the most common t8;14 in BL, variant chromosomal abnormalities are also found which involve the c-*myc* locus and the other two immunoglobulin gene loci [t2;8 involving kappa light (L) chains at 2p12 and t8;22 involving lambda L-chains at 22q11]. Thus the ability of the immunoglobulin genes to rearrange seems to be a crucial feature of these genes, causing their erroneous involvement in chromosomal translocations.

The more recent molecular analysis of the T cell receptor (TCR) has shown that the genes which encode these proteins also undergo DNA rearrangement during normal T cell development (reviewed in Hood et al. 1985). By analogy with the tumour-associated immunoglobulin-"mediated" translocations discussed above, it seemed likely that the TCR genes would also be involved in tumour-specific chromosomal abnormalities in T cell leukaemias. The TCR α chain genes in humans have been mapped to 14q11 (Caccia et al. 1985, Croce et al. 1985), a site which coincides with one of the breakpoints seen in the T cell chronic lymphocytic marker chromosome inversion (inv)14 (q11;q32) (Aurias et al. 1980, Zech et al. 1984). We have recently cloned the junction of one of these inversion chromosomes and we found that an immunoglobulin H-chain variable (V) region gene and TCR joining (J) α segments had fused as a result (Baer et al. 1985). In this paper, we discuss the possible relevance of this fusion gene to the development of some T cell leukaemias in man.

Results

Inv14 Results in an IgT Gene in SUP-T1 Cells

The junction of one of the breakpoints from an inv14 chromosome present in SUP-T1 cells [derived from a T cell lymphoma (Hecht et al. 1984)] has recently been cloned and we found that an immunoglobulin (Ig)/TCR

[1]Medical Research Council, Laboratory of Molecular Biology, Hills Road, Cambridge, CB2 2 QH, U.K.

37. Colloquium Mosbach 1986
Cell Cycle and Oncogenes

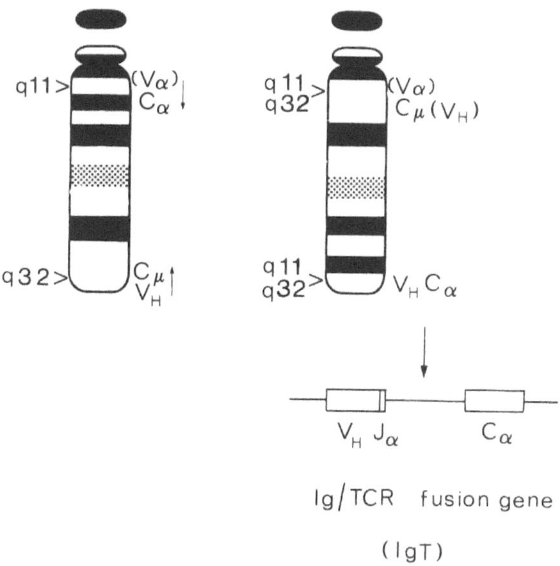

Fig. 1. Organisation of TCR and immunoglobulin genes on normal and inverted chromosome 14 in a human T cell lymphoma, SUP-T1. The diagram illustrates banded homologous chromosome 14 from a cell line carrying inv14. Location of V and C regions of the various genes are shown together with the location and structure of the IgT gene resulting from the inversion of chromosome 14

fusion gene (IgT) had been formed at 14q32 (Baer et al. 1985). Production of this fusion gene essentially mimics the normal process of DNA rearrangement which takes place within the TCRα or Ig loci during T or B cell differentiation respectively. These loci, in the unrearranged, undifferentiated state, have separate V, diversity (D), J and C region genes and the V, D and J segments undergo a joining process to create continuous V-D-J region. The process of V-D-J joining is however fairly imprecise so that aberrant or non-productive rearrangements can result which will not produce functional protein. Frequently individual B or T cells carry one productively and one non-productively rearranged allelic chromosome of the rearranging genes. The IgT fusion gene (illustrated in Fig. 1) occurs after the piece of chromosome 14 carrying joining segments of the TCR α locus together with the TCR Cα gene (normally at 14q11) has flipped over, bringing Jα-Cα nearer to the telomere, at 14q32. The breakpoint at 14q32 seems to have occurred within the Ig H-chain V-genes allowing a join to occur betwen an IgV_H segment, probably an α-chain D segment, and a Jα segment. The resulting IgT gene is productively rearranged, which is to say that the V-D and D-J junctions are in-frame with each other to allow a putative mature V+C region mRNA to be produced (Baer et al. 1985). The homologous chromosomes (i.e., the normal chromosome 14) in SUP-T1 cells has a non-productive rearrangement of a TCR Vα with a Jα segment; this chromosome is, therefore, unlikely to produce functional TCR α chain protein.

Expression of IgT in SUP-T1

The observation that an IgT fusion gene results from the inversion 14 chromosomal abnormality in the T cell lymphoma studied suggests that the putative resultant protein molecule might have a role in leukaemogenesis; speculations about how this might occur are presented in the discussion. Presumably, however, a role in oncogenesis necessitates expression of the IgT gene. We have, therefore, analysed mRNA prepared

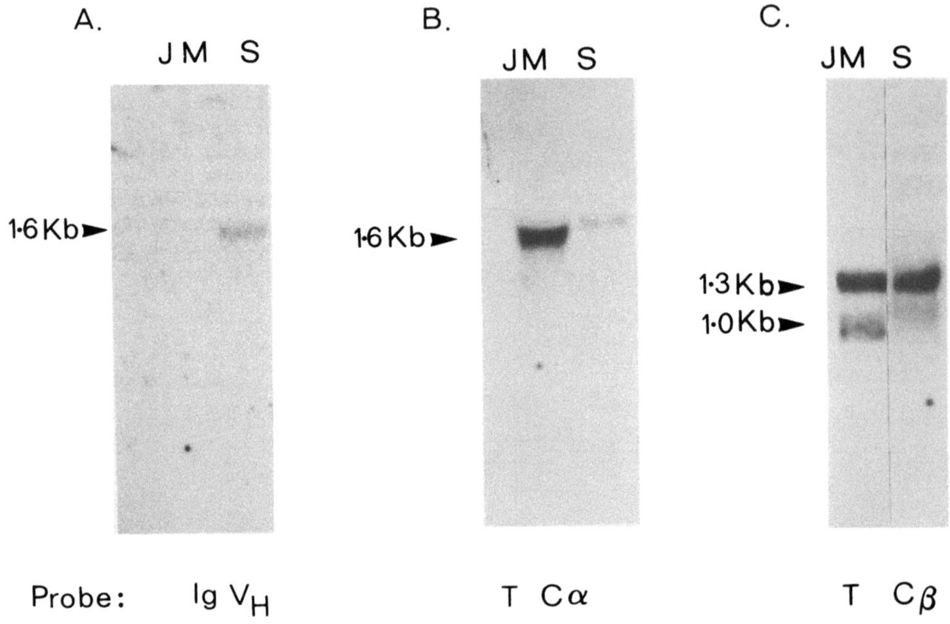

Fig. 2. RNA hybridisation analysis of mRNA prepared from T cell leukaemia cell lines. RNA was prepared from either JM or SUP-T1 cells as described (Favaloro et al. 1980). Poly(A) mRNA was selected on oligo dT, 5 µg amounts glyoxylated (Thomas 1980) fractionated on 1.4% agarose and transferred to Hybond (Amersham Int.). Filters were hybridised as previously described (Sims et al. 1984) to either (A) an immunoglobulin V_H probe (Baer et al. 1985), (B) to a TCR Cα probe or (C) to a TCR Cβ probe (Sims et al. 1984). After hybridisation at 65°C in 6 x SSC, the filters were washed with 0.1 x SSC prior to autoradiography at -70°C with prefogged film

from SUP-T1 for transcripts of the IgT fusion gene, and the results of this analysis are shown in Fig. 2. When RNA from SUP-T1 and a T cell line called JM (which does not have abnormalities of chromosome 14 (J. Walters and THR unpublished observations) were compared in Northern filter hybridisation using a TCR Cβ probe, we found β-chain transcripts originating from both alleles in SUP-T1 and in JM (Fig. 2A). The upper hybridising bands (1.3 kb) probably come from the productively rearranged alleles, whereas the smaller molecular weight, fainter band (about 1 kb) probably comes from the aberrantly rearranged Cβ homologues which exists in both JM (Collins et al. 1985) and in SUP-T1 cells (LMG and THR, unpublished observations). In addition, T3 mRNA (δ chain) has been shown to be synthesised in both cell lines (Tunnacliffe et al. 1986). A separate filter carrying both SUP-T1 and JM mRNA was hybridised to a probe which derived from the V_H gene rearranged at the inv14 junction in SUP-T1. The probe detected a hybridising species in SUP-T1 RNA and none in JM (Fig. 2B) consistent with the transcription of IgT in SUP-T1 cells. To assess whether this hybridisation component also carries TCR Cα sequences, the filter was rehybridised with a Cα cDNA probe after the V_H hybridising probe was effectively removed from the filter. The Cα probe detected the same band in SUP-T1 RNA and in addition a band of approximately the same mobility in JM RNA (Fig. 2C). This supports the idea that an mRNA is being produced from the IgT gene in SUP-T1 cells; the relative quantity of the IgT mRNA seems to be approximately 10% of the TCRα mRNA in JM.

```
                  Vβ                                 Dβ
 S  S  L  E  L  G  D  S  A  L  Y  F  (C) A  S  S  V  G  G  S
GAGCTCTCTGGAGCTGGGGGACTCAGCTTTGTATTTCTGTGCCAGCAGCGTAGGTGGTAG
         10        20        30        40        50        60

             Jβ2.1
 L  K  Q  F  F  G  P  G  T  R  L  T  V  L
CTTGAAACAGTTCTTCGGGCCAGGGACACGGCTCACCGTGCTAGGTAAGAAGGGGGCTCC
         70        80        90       100       110       120
```

Fig. 3. Nucelotide sequence of the productively rearranged allele of TCR β-chain in SUP-T1 cells. λ phage clones containing the TCR β-gene were isolated from a library of SUP-T1 genomic DNA (Baer et al. 1985) using a Cβ cDNA probe (Sims et al. 1984). Both β-chain alleles in SUP-T1 are rearranged at the Cβ2 locus, so the precise location of rearrangements were identified within the Jβ2 region by comparison of the clone maps with those of the germ-line. The nucleotide sequence of the site of V-J joining on one allele was determined in M13 vectors as described (Bankier and Barrell 1983, Sanger et al. 1980). Only the sequence at the V-D-J junction of the productive allele is illustrated

Production of IgT mRNA in SUP-T1 cells is obviously a prerequisite for protein production at the cell surface. Further, if the IgT protein were to interact in a conventional way to form a heterodimeric receptor structure, a β-chain protein must also be produced by the cells. The RNA filter hybridisation experiment shown in Figure 2 demonstrates the existence of full length β-chain mRNA in SUP-T1 but in order to confirm the occurrence a productive β-allele, we isolated the rearranged Cβ-genes from SUP-T1 DNA. The nucleotide sequence of the rearranged Vβ genes was then determined. Southern filter hybridisation showed that rearrangement had occurred to both Cβ2 genes in SUP-T1 cells (data not shown). The sites of rearrangement within λ phage clones corresponding with the two Cβ alleles were determined by comparison to the genomic sequence of the unrearranged Jβ2 locus (Tunnacliffe and Rabbitts 1985). One set of clones possessed a rearrangement at the Jβ2.1 segment and analysis of the predicted coding sequences showed that a Vβ gene had joined at this position. Figure 3 shows a partial nucleotide sequence of this V-D-J junction. The Vβ gene has joined productively and apparently a Dβ segment is also involved. The full size β-chain mRNA detected in the Northern filter hybridisation shown in Fig. 2, therefore, probably represents the transcription product of this productively rearranged β-chain gene. Thus the SUP-T1 cells make mRNA for both IgT and TCR β chains.

The putative protein products made from mRNAs are presumably capable of combining at the surface of the SUP-T1 cells as a heterodimer (in which disulphide bonding would take place between the Cα and Cβ domains). Experiments are in progress to assess the presence of heterodimer on SUP-T1 cell surface. Indirect evidence for receptor expression in SUP-T1 however comes from the detection of T3 antigen on the surface of these cells (Smith et al. 1984).

Discussion

Is IgT Relevant to T Cell Leukaemia?

Our investigation of the inv14 chromosome abnormality in SUP-T1 cells and of the expression from the resultant IgT locus suggests that a chimaeric receptor, made up from the IgT molecule (Ig V_H + TCR Cα) and the normal β-chain molecule (TCR Vβ + Cβ), may be produced as a consequence. The expression of surface T3 antigen supports this conclusion since it is thought this antigen only appears on the cell surface in association with receptor (Weiss and Stobo 1984). The involvement of IgT in this putative receptor, rather than normal α-chain, is inferred from the fact that the homologous chromosome in SUP-T1 cells has a non-productive α-chain rearrangement which cannot yield α-chains. The possibility of a β-β homodimer associated with T3 on the cell surface, of course, cannot be ruled out. The question posed by these observations is whether such a chimaeric receptor has any relevance to oncogenesis in cells with inv14.

Two major abnormalities involving the long arm of chromosome 14 have been described which bear on this discussion viz. inv14 and t14;14. In both of these, the breakpoints are q11 and q32. There is a correlation of inv14 to T-CLL (Zech et al. 1984) although the numbers of cases analysed is still low. However, the correlation of inv14 to these T cell tumours suggests, like the CML (Ph' positive) and BL translocations, a role for this abnormality in leukaemiogenesis. If this is the case, is IgT the mediator? At present we have only been able to clone one inv14 junction so we cannot yet assess whether IgT is a common feature of inv14 marker chromosomes. Cloning experiments designed to identify other inv14 junctions within the Jα locus are underway but are hampered by the very large size of this genomic region (Arden et al. 1985, Hayday et al. 1985, Winoto et al. 1985, Yoshikai et al. 1985). It will clearly be necessary to clone several inv14 junctions to see whether IgT is a common feature, to analyse sequence relationships between the putative V_H segments joined to the TCR Jα and also to analyse sequence relationships of TCR Vβ segments which would also contribute to the "combining site" of putative receptor molecules.

An important aspect of the IgT gene in SUP-T1 is that it was created from the rearrangement of V, D and J-segments and that the overall joining processes resulted in a productive gene. The probability of generating a functional IgT gene is low becauce, in addition to the inversion process itself, V-D and D-J joins can occur out of frame, particularly when N-region diversity is taken into account. The observation that a productive IgT gene can be generated supports, at least superficially, the contention that IgT is an important feature of T-cell leukaemogenesis in those cases carrying inv14. However, it is possible that IgT is an important molecule for the normal immune system (e.g., as a means to expand the T-cell receptor repertoire) and that IgT formation is simply a natural, harmless consequence of inversion of chromosome 14. If the latter is true, it should be possible to find cells in which inv14 is associated with a non-productive IgT arrangement and a productive TCR α rearrangement on the allelic chromosome. Recently, inv14 chromosomes have been noted in in vitro T-cell proliferations at a very low incidence (Aurias et al. 1985) but the molecular organization of these inv14 chromosomes is not known.

IgT Chimaeric Receptor – Possible Mechanisms of Auto-Triggering in T Cell Leukaemias

If IgT is important to leukaemogenesis, how could the protein product cause cell proliferation? The function of normal T cell receptor is to bind antigen and thereby transmit a signal which activates the T cell to respond and initiate cell division. The IgT receptor, in which V_H replaces Vα, may cause an activation response which is independent of antigen. This auto-activation might be caused by a different folding of the V_H domain compared with the $V\alpha$ domain; the overall "combining site" might thereby transmit signal directly to the inside of the cell. Alternatively, the IgT receptor might recognise a self-antigen which could cause triggering.

If any of these mechanisms are operative, it means that inv14 could be the first event in a number of steps towards overt T cell leukaemia. Thus the pre-leukaemic cell may undergo an inversion of chromosome 14 which may or may not produce IgT. If productive, the T cell has the chance to become activated and undergo a response which involves cell division over a prolonged period of time. During this time (which may or may not be finite), the preleukaemic cell has the chance to acquire further abnormalities (e.g., such as involving proto-oncogenes) which may themselves generate the truly leukaemic state. Therefore, in this model IgT would be regarded as an initiator of cell activation which, because it is "self-driven", immortalises the T cell for a protracted time during which secondary oncogenic events can occur. Biological systems for the analysis for IgT will be needed to assess these various features and alternatively to assess whether IgT can operate as part of the normal immune system. If IgT is not a common feature of inv14, then presumably some other feature plays this initiator role.

References

Arden B, Klotz JL, Siu G, Hood LE (1985) Diversity and structure of genes of the α family of mouse T cell antigen receptor. Nature (London) 316:783-787

Aurias A, Dutrillaux B, Buriot D, Lejeune J (1980) High frequencies of inversions and translocations of chromosomes 7 and 14 in ataxia telangiectasia. Mutat Res 69:369-374

Aurias A, Couturier J, Dutrillaux AM, Dutrillaux B, Herpin F, Lamoliatte E, Lombard M, Muleris M, Paravatou M, Prieur M, Prod'homme M, Sportes M, Viegas-Pequignot E, Volobouev V (1985) Inversion 14; the most frequently acquired rearrangement in lymphocytes. Hum Genet 71:19-21

Baer R, Stinson MA, Forster A, Smith SD, Rabbitts TH (1985) Fusion of an immunoglobulin variable gene and a T cell receptor constant gene in the chromosome 14 inversion associated with T cell tumours. Cell 43:705-713

Bankier AT, Barrell BG (1983) Shotgun DNA sequencing. Nucleic Acid Biochem B508:1-34

Caccia W, Bruns GAP, Kirsch IR, Hollis GF, Bartness V, Mak TW (1985) T cell receptor α chain genes located on chromosome 14 at 14q11-14q12 in humans. J Exp Med 161: 1255-1260

Collins MKL, Kissonerghis AM, Dunne MJ, Watson CJ, Rigby PWJ, Owen MJ (1985) Transcripts from an aberrantly rearranged human T cell receptor β chain gene. EMBO J 4:1211-1215

Croce CM, Isobe M, Palumbo A, Puck J, Ming J, Tweardy D, Erikson J, Davis M, Rovera G, Rovera G (1985) Gene for α chain of human T cell receptor: location of chromosome 14 region involved in T cell neoplasms. Science 227:1044-1047

Favaloro J, Treisman R, Kamen R (1980) Transcriptional maps of polyoma virus-specific RNA: Analysis by 2-D nuclease S_1 gel mapping. In: Grossman L, Moldave K (eds) Methods in enzymology, vol 68. Academic Press, London New York, pp 718-749

Hayday AC, Diamond DJ, Tanigawa G, Heilig JS, Folsom V, Saito H, Tonegawa S (1985) Unusual organisation and diversity of T cell receptor α chain genes. Nature (London) 316:828-832

Hecht F, Morgan R, Kaiser-McCaw Hecht B, Smith SD (1984) Common region on chromosome 14 in T-cell leukaemia and lymphoma. Science 226:1445-1447

Hood L, Kronenberg M, Hunkapiller T (1985) T-cell antigen receptors and the immunoglobulin supergene family. Cell 40:225-229

Rabbitts TH (1985) The c-myc protooncogene: involvement in chromosomal abnormalities. Trends Genet 1:327-331

Sanger F Coulson AR, Barrell BG, Smith AJH, Roe BA (1980) Cloning in single-stranded bacteriophage as an aid to rapid DNA sequencing. J Mol Biol 143:161-178

Sims JE, Tunnacliffe A, Smith WJ, Rabbitts TH (1984) The complexity of human T cell receptor β chain variable and constant region gene. Nature (London) 312:541-545

Smith SD, Shatsky M, Cohen PS, Warnke R, Link MP, Glader BE (1984) Monoclonal antibody and enzymatic profiles of human malignant T lymphoid cells and derived cell lines. Cancer Res 44:5657-5660

Southern EM (1975) Detection of specific sequences among DNA fragments separated by gel electrophoresis. J Mol Biol 98:503-517

Thomas PS (1980) Hybridisation of denatured RNA and small DNA fragments transferred to nitrocellulose. Proc Natl Acad Sci USA 77:5201-5205

Tunnacliffe A, Rabbitts TH (1985) Sequence of the Dβ2-Jβ2 region of the human T cell receptor β-chain locus. Nucleic Acids Res 13:6551-6661

Tunnacliffe A, Sims JE, Rabbitts TH (1986) T3δ pre-mRNA is transcribed from a non-TATA promoter and is alternatively spliced in human T cells. EMBO J 5:1245-1252

Vieira J, Messing J (1982) The pUC plasmids, an M13mp7-derived system for insertion mutagenesis and sequencing wiht synthetic universal promoters. Gene 19:259-268

Weiss A, Stobo JD (1984) Requirement for the co-expression of T3 and the T cell antigen receptor on a malignant human T cell line. J Exp Med 160:1284-1299

Winoto A, Mjolsness S, Hood L (1985) Genomic organization of the genes encoding mouse T cell receptor α chain. Nature (London) 316:832-836

Yoshikai Y, Clark SP, Taylor S, Sohn U, Wilson BI, Minden MD, Mak TW (1985) Organisation and sequences of the variable, joining and constant region genes of the human T cell receptor α chain. Nature (London) 316:837-840

Zech L, Gahrton G, Hammerström L, Juliusson G, Mellstedt H, Robert KH, Smith CIE (1984) Inversion of chromosome 14 marks human T cell chronic lymphocytic leukaemia. Nature (London) 308:858-860

Regulation of c-*myc* in Variant Chromosomal Translocations of Burkitt Lymphoma

M. Lipp[1], R. Schilling[1], S. Wiest[2], and P. Hartl[1]

Introduction

The c-*myc* gene, the cellular counterpart of the oncogene of the avian retrovirus MC29 (Duesberg et al. 1980) appears to play an important role in normal cell proliferation. In response to exogenous stimuli like platelet-derived growth factor (PDGF) in fibroblasts and lectin or anti-IgM in lymphocytes, c-*myc* mRNA is accumulated in the cytoplasm about 2-3 h after induction, prior to the onset of cellular DNA synthesis (Armelin et al. 1984, Kelly et al. 1983, McCormack et al. 1984). The c- *myc* gene is composed of three exons, the first of which is noncoding (Battey et al. 1983, Bernard et al. 1983, Hamlyn and Rabbitts 1983, Watt et al. 1983, Gazin et al. 1984). Transcription of the c- *myc* gene proceeds from two initiation sites 160 bp apart from each other at the beginning of the first exon. However, two proteins of 64 and 67 K, located in the nucleus were shown to be coded for by the c-*myc* gene. Their function is not yet understood (Hann and Eisenman 1984, Persson et al. 1984, Persson and Leder 1984).

Perturbations of the c- *myc* gene have been shown to occur in many B-cell malignancies in chicken, mice, and man. One of the best-analyzed examples are human Burkitt lymphomas and murine plasmacytomas carrying consistently a reciprocal translocation [t(8;14), t(2;8), and t (8;22) in man, and t(12;15), and (t(6;15) in mice] which always involves one of the immunoglobulin heavy or light chain genes and the locus of the c- *myc* oncogene on human chromosome 8 and on mouse chromosome 15 (for review see Klein 1983, Klein and Klein 1985). In many cases of Burkitt lymphomas with the common t(8;14) translocation which involve the heavy chain locus, the chromosomal translocations disrupt the first noncoding exon or the first intron of th c- *myc* gene, and thus separate the body of the gene from its physiological promoters, possibly causing deregulated transcription of the c- *myc* gene from otherwise cryptic promoters (Adams et al. 1983, Ar-Rushdi et al. 1983, Bernard et al. 1983, Hayday et al. 1984, Taub et al. 1984). In cases with the variant t(2;8) and t(8;22) translocations, which appear each in about 10% of all patients, and involve the kappa or lambda light chain genes, the breakpoint is usually far downstream of the c-*myc* gene, thus leaving the c-*myc* transcription unit intact (Croce et al. 1983, Davis et al. 1984, Erikson et al. 1983, Rappold et al. 1984).

Multiple somatic mutations in the first exon have been described in four cases in which the breakpoint does not disturb the c- *myc* gene (Rabbitts et al. 1983, 1984, Taub et al. 1983, Denny et al. 1985). It is not known whether generation of somatic mutations is a general mechanism leading to a deregulation of the c- *myc* gene in cells with

[1]Institut für Biochemie der Universität München, Karlstraße 23, D-8000 München 2, FRG

[2]Institut für Virologie, Zentrum für Hygiene der Universität, Hermann-Herder-Str. 11, D-7800 Freiburg, FRG

37. Colloquium Mosbach 1986
Cell Cycle and Oncogenes

variant translocations. The analysis of the c-*myc* mRNA in Burkitt lymphomas demonstrates that only the c- *myc* gene of the translocated chromosomes in transcribed, while the normal allele remains silent (Ar-Rushdi et al. 1983, Bernard et al. 1983, Nishikura et al. 1983, Taub et al. 1983). Mechanisms operating at a posttranscriptional level also seem to play an important role in the regulation of the c-*myc* gene (Dani et al. 1984, Piechaczyk et al. 1985). The target structures for transcriptional and posttranscriptional regulation of the c-*myc* gene have not yet been identified.

We have molecularly analyzed the reciprocal chromosomal breakpoints as well as both c- *myc* loci of the Burkitt lymphoma (BL) line BL64 in which translocation joins the immunoglobulin kappa locus on chromosome 2 to the c- *myc* gene on chromosome 8. One important finding of this characterization was the detection of recognition sequences that apparently mediate site-specific recombination during immunoglobulin rearrangement in the vicinity of both breakpoints (Lipp and Hartl 1986). In order to determine the parts of the c-*myc* gene required for transcriptional and posttranscriptional regulation, and to detect possible functional differences between the translocated and nontranslocated allele we have constructed hybrid molecules containing part of the c-*myc* promoter-leader region and the bacterial chloramphenicol acetyltransferase (CAT) gene. The activity of the fused genes was tested after transfection into various cell types (Lipp et al. 1987).

Analysis of the t(2;8) Translocation in BL64

In the variant t(2;8) translocations of Burkitt lymphoma determination of the breakpoint based so far mainly on data obtained by in situ hybridization on metaphase chromosomes, revealed that the c-*myc* gene was retained on the $8q^+$ chromosome, and that a region 5' of the constant (C_k), within the joining (J_k), or the variable (V_k) region may be affected (Erikson et al. 1983, Emanuel et al. 1984, Rappold et al. 1984, Taub et al. 1984). The results obtained by Southern blot analysis of genomic BL64 DNA confirmed those of the in situ hybridizations (Hameister and Adolph 1986), and strongly suggested that the breakpoint of the translocation in BL64 is located within the joining region. To analyze in more detail the structural changes caused by the translocation, we screened a genomic library of BL64 with a J_k probe, and isolated recombinant lambda phages containing the functional rearranged kappa light chain gene, or the reciprocal breakpoints on chromosome $8q^+$ and chromosome $2p^-$. Detailed mapping experiments, together with the analysis of the DNA sequences spanning the breakpoints on the two marker chromosomes revealed that the chromosomal breaks occurred on chromosome 2 5' of the J_5-segment within the conserved nonamer and heptamer recombination sequences, and on chromosome 8 within a cluster of highly repetitive Alu-type sequences (Lipp and Hartl 1986).

Genesis of the Translocation by Site-Specific Recombination

Our finding that one of the chromosomal breaks has occurred between the putative V-J recombination signal caused us to look for similar target sequences, usually found 3' of non-rearranged V_k-genes, in the vicinity of the breakpoint of chromosome 8. Both signal boxes could be found as inverted repeats of the J-sequences. The nonamer 5'-A/CA-AAAATT-3' corresponds to 5'-GATTTTTG/T-3' of J_k and includes the breakpoint (/) observed in chromosome $8q^+$. The heptamer 5'-CACAGTG-3' corresponds to 5'-CACTGTG-3' of J_k and is located at a distance of 155 bp upstream of the nonamer box, within the unique sequences of chromosome 8. These data strongly suggested that enzymes normally catalyzing V-J rearrangement could be involved in the generation of this t (2;8) translocation.

The presence on both chromosomal partners of a complete signal structure consisting of a heptamer and a A/T-rich nonamer, which appear to be highly conserved between all immunoglobulin and T-cell receptor gene segments (for review see Tonegawa 1983, Hood et al. 1985), led us to propose a secondary structure for the alignment of chromosome 2 and 8 (Fig. 1) based on the suggestions for V-J- or V-D-J-joining (Seidman et al. 1979, Tonegawa 1983). We suggest that the putative recombinases can recognize both sequences and stabilize the whole interchromosomal recombination structure. This interpretation of our results is further strengthened by a recently reported nuclear endonucleolytic activity that cleaves immunoglobulin recombination sequences (Hope et al. 1986). This activity introduces a double-strand cut, and the cleavage occurs predominantly at the dinucleotide pair $\frac{\text{TG}}{\text{AC}}$. The generation of the breakpoint on chromosome $8q^{+}$ could be explained by the action of such an enzyme introducing a double-strand cut (/) in the stem structure of the paired nonamers $\frac{\text{GATTTTTG/T}}{\text{TTAAAAAC/A}}$ followed by ligation of the nucleotides 5'-A-T-3' (Fig. 1). Such a physical linkage of chromosome 8 to 2 could have been the crucial step for the initiation of the translocation.

Additional evidence for an involvement of the immunoglobulin recombinase system in the genesis of chromosomal rearrangements was recently found in a t(11;14) translocation of chronic lymphocytic leukemia, where a weak homology to the J_H recombination sequences could be detected (Tsujimoto et al. 1985). However, in cases of the common t(8;14) translocations of BL, where the breakpoint is often localized within the Sμ region of the heavy chain, no such homology could be found. Possibly this is a special feature of the variant t(2;8) translocations. However, the perfect homology of the recombination target structures observed in BL64 could be an exception. Possibly, a weaker homology between a transcriptionally active site in the chromatin and the immu-

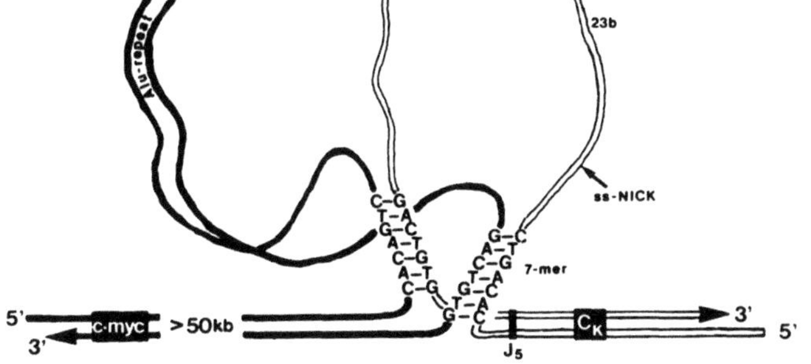

Fig. 1. Possible secondary structure for the alignment of chromosome 2 and 8 via the heptamer and nonamer recognition sequences. *Arrows* indicate the locations of the reciprocal breakponts, and possible target sites for the putative recombinase system (for explanation see text). The *c-myc* gene is located at least 50 kb upstream of the breakpoint in the J_k-region

nolobulin loci could be sufficient to transiently stabilize an interchromosomal complex and to activate some of the components of the immunoglobulin recombination system leading to the initiation and an irreversible manifestation of a particular translocation.

Molecular Analysis of the c-*myc* Loci

In order to get some information about the distance of the translocated c-*myc* locus from the breakpoint in the J-region, and about possible structural alterations, we isolated c-*myc*-specific lambda phages from the genomic library of BL64, and performed genomic walking experiments starting from the 3'-part of the c-*myc* gene. As indicated in Fig. 1, we have analyzed a region of about 50 kb downstream of the c-*myc* gene with no linkage to the kappa locus so far. Thus, if the translocation event is responsible for the activation of the c-*myc* gene, a special mechanism which influences the expression of a gene over such long distances must be assumed.

Constitutional Gene Duplication in one c-*myc* Allele

In contrast to other variant t(2;8) translocations, we detected by Southern blot analysis of Bl64 DNA a rearrangement of the c-*myc* locus. The structure of this alteration was identified by detailed mapping of overlapping c-*myc*-specific lambda clones isolated from the BL64 library. As summarized in Fig. 2, one allele has a normal configuration, whereas the other allele shows an alteration consisting of a duplication of a 9.0 kb fragment including the two coding exons. Exactly the same changes in the hybridization pattern were detected with DNA isolated from the lymphoblastoid cell line, IARC 549, established from peripheral blood leukocytes of the same patient. Hence, we concluded that the observed alteration of the c-*myc* gene in both cell lines is a constitutional restriction fragment length polymorphism (RFLP) probably unrelated to the translocation event.

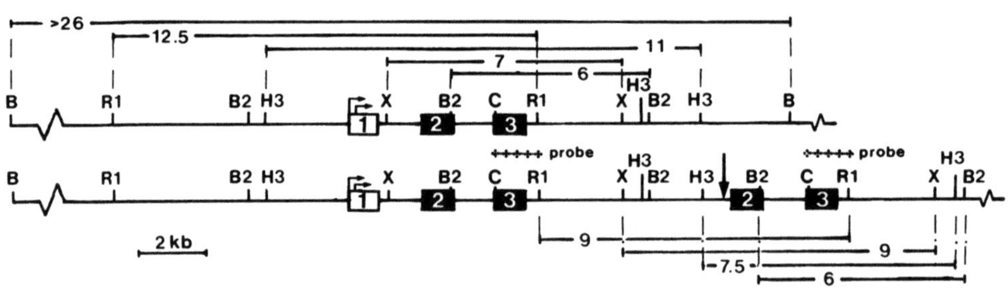

Fig. 2. Diagrammatic representation of the structure of both c-*myc* alleles in the BL64 cell line deduced from detailed restriction enzyme analysis of overlapping lambda clones isolated from a genomic library. *Blocks* designate possible transcribed regions, *filled blocks* indicate the coding sequences of the expressed gene, and *arrows* show the normal transcriptional start sites. The length and the location of normal (*above*) and of additional single copy fragments (*below*) hybridizing with a 3'-c-*myc*-specific probe are indicated. The vertical arrow shows the breakpoint of the 9.0 kb duplication. For abbreviations of restriction enzymes see Fig. 4

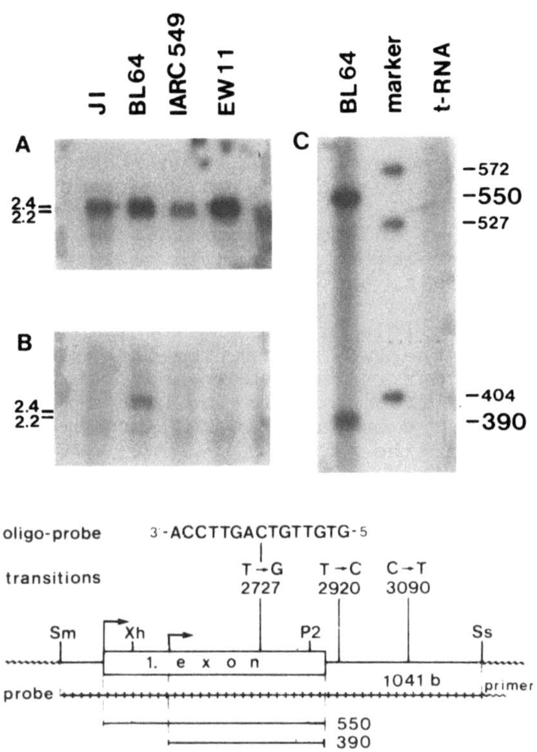

Fig. 3. Northern (A,B) and S1 (C) analysis of c-*myc*-specific RNA expressed in BL64 cells. The same Northern blot hybridized with a 3'-probe (A) was rehybridized with the allele-specific oligonucleotide probe spanning the point mutation found in exon one (B and bottom panel). In BL64, as in most other BL's, the first promoter is used preferentially (C). The probe used for the S1 analysis, and the location of protected fragments are indicated in the bottom panel

Point Mutations in the Expressed c-*myc* Allele

We wished to determine whether the c- *myc* allele carrying the duplication is translocated and expressed, or not. This knowledge would allow us to identify functional differences in that allele involved in the translocation. Due to the isolation of lambda clones, which contain the first noncoding exon as well as sequences from the second exon of the duplication, we were able to distinguish between both alleles of the c-*myc* locus and to compare the 1041 SmaI-SstI fragment of both alleles containing the first exon and 5'- and 3'-flanking regions by sequencing. We detected, as shown in Fig. 3, three point mutations only in the sequence originating from the allele without the duplication.

In order to establish if the transcription pattern was influenced by the duplication, we performed Northern-, and S1-analysis with total RNA isolated from BL64 cells. As shown in Fig. 3A and C, the c- *myc* gene is expressed strongly in BL64 cells, and the mRNA's initiated at the two possible capsites of the dual c-*myc* promoter. In BL64, as in most other Burkitt lymphomas, the first promoter is used preferentially. These results suggested that the allele carrying the mutations rather than the allele with the duplication might be involved in the translocation and therefore expressed. This suggestion was tested by using a synthetic oligonucleotide of 15 bases spanning the point mutation found in position 2727 of exon one as an allele-specific hybridization probe. Rehybridization of the Northern blot already shown in Fig. 3A with the oligonucleotide probe revealed that only c-*myc* RNA from BL64 cells, but not those from the other cells hybridized with this oligonucleotide (Fig. 3B). Thus we concluded that only the allele carrying the mutations is expressed in BL64 cells.

Activity of the Dual c- *myc* Promoter

The somatic mutations observed in the first exon of BL64 and other BL's suggest that this long untranslated leader region carries important signals for the regulation of the c-*myc* gene. In order to identify target sequences required for transcriptional and posttranscriptional regulation we have constructed hybrid molecules containing three c-*myc* promoter-leader regions of different origin, and the bacterial chloramphenicol acetyltransferase gene (CAT). One c- *myc* promoter region was cloned from the normal, nontranslocated allele of the Burkitt lymphoma line Raji (Rappold et al. 1984), and the two others from both alleles of the BL64 line. The Raji-derived constructs used are schematically described in Fig. 4. The 862 bp PvuII fragment carrying both c-*myc* promoters and most of the noncoding first exon was selected, and subcloned in the plasmid pUC-CAT1 generating the clone pmyc-CAT-P. Deletion constructs from pmycCAT-P were obtained by removing upstream sequences to generate the clones p *myc*CAT-S and p*myc* CAT-X. The first promotor was selectively removed from p*myc* CAT-P to generate p *myc*CAT-PΔP1. The second promoter was selectively removed downstream of the XhoI site to generate p *myc*CAT-PΔP2. Similarly, 302 bp of the 5'-end of the first exon were deleted from p *myc*CAT-P(b) by digestion with NaeI and XbaI to generate p*myc* CAT-PΔN. Upstream sequences were reinserted in p*myc* CAT-P to generate p *myc*CAT-R1.

Activity in Various Cell Types. The activity of the c- *myc* promoter was analyzed in titration experiments in which the amount of p*myc* CAT-P DNA, used for transient transfection, was plotted against the resulting CAT activity. The SV40 promoter in front of the CAT gene (Gorman et al. 1982) was used as an internal standard. As shown in Fig. 5, p*myc*-CAT-P and pSV2CAT have quite different titration curves. The activity of the two promoters was compared quantitatively by comparing the CAT activities mediated by equimolar nonsaturating amounts of DNA and also by comparing the amounts of DNA required for half maximal CAT activity. From these data it becomes apparent that the c- *myc* promoter is 15-fold, 20-fold, and 30-fold less active than the SV40 promoter in baby hamster kidney cells (BHK), mouse L cells, and the human permanent fibroblast line 143tk$^-$, respectively. The importance of making titration curves for a quantitative comparison of promoter strength is clearly illustrated in Fig. 5, since transfection of large amounts (10 to 20 μg) of p*myc* CAT-P and pSV2CAT resulted in about the same final CAT activities in BHK and L cells despite the described 15- and 20-fold difference in the activity of both promoters.

Different Activities of Both Promoters. In an attempt to identify the sequences required for activity and regulation of the c-*myc* promoter we have tested the activities of the various deletion constructs by titration analysis in L cells (Fig. 6A). p *myc* CAT-S with 248 bp deleted upstream of both promoters exhibited a lower activity than p*myc* CAT-P. Removal of the second promoter and the noncoding sequences downstream from p*myc* CAT-P generating the construct p*myc* CAT-PΔP2, decreased the activity more dramatically (five- to six-fold). In contrast, removal of the 175 bp SmaI-XhoI fragment carrying the first promoter (p*myc* CAT-PΔP1) increased the resulting CAT activity. Comparison of the constructs lacking the first or the second promoter, revealed an at least six- to eight-fold higher activity of the plasmid carrying the second promoter. The activity of this promoter is similar to that of the plasmid pAd12-1aCAT (Kruczek and Doerfler 1983) containing the Ad12 E1a-promoter which is expected to carry an enhancer element, as already could be

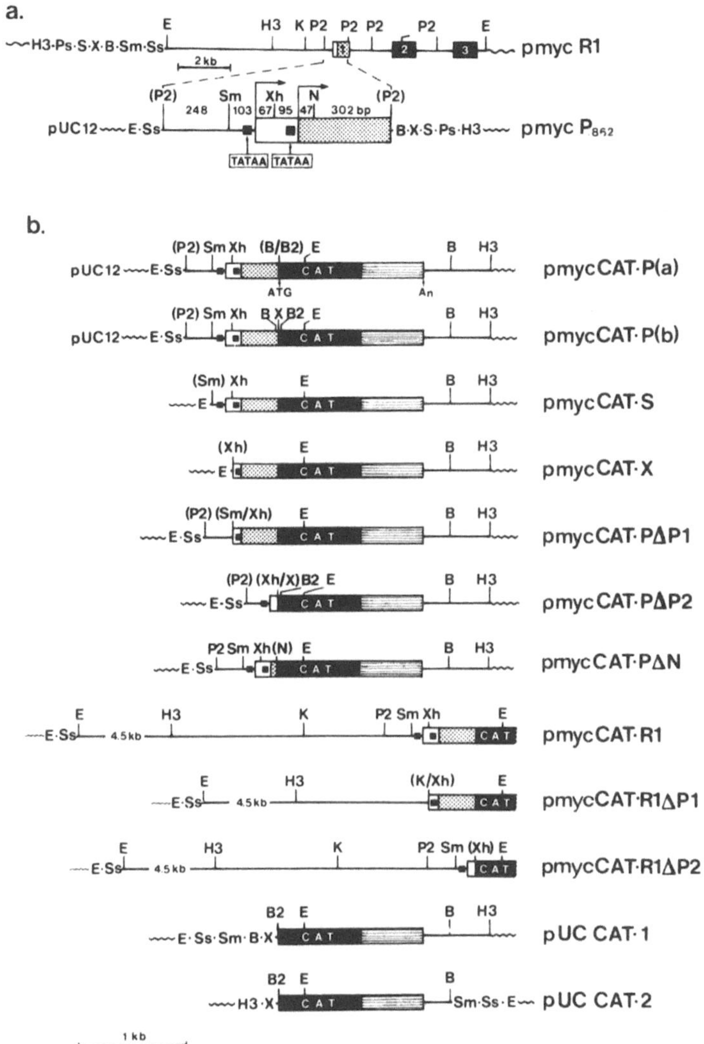

Fig. 4. Schematic representation of constructed hybrid molecules of the human c-*myc* and the bacterial chloramphenicol acetyltransferase gene (CAT). *Blocks* designate possible transcribed regions, *filled blocks* indicate the coding sequences of the expressed gene, *arrows* show the normal transcriptional start sites, *lined blocks* indicate the SV40-derived splice and polyadenylation site, and *filled boxes* designate the presumptive Hogness-Goldberg signals (TATAA). (a) The 12.5 kb EcoRI fragment carrying the human c-*myc* gene was cloned from the normal allele of the Burkitt lymphoma line Raji in the phage EMBL4 and subcloned into the EcoRI site of pUC12 (pmycR1). The 862 bp PvuII fragment containing both c-*myc* promoters and most of the noncoding fist exon was subcloned in the SmaI site of pUC12 (pmycP$_{862}$). The dotted block designates the transcribed region, if the second promoter is used, and the open block the additional transcribed portion if the first cap-site is used. (b) Hybrid molecules of the human c-*myc* 862 bp PvuII fragment and the bacterial CAT gene (p*myc*CAT-P), and constructions derived from p*myc*CAT-P. For description see text. *Wavy lines* designate pUC12 sequences. Abbreviations of the restriction enzymes: *B* BamHI; *B2* BglII; *E* or *R1* EcoRI; *H3* HindIII; *K* KpnI; *N* NaeI; *Ps* PstI; *P2* PvuII; *S* SalI; *Sm* SmaI; *Ss* SstI; *X* XbaI; and *Xh* XhoI. Restriction sites in parentheses are lost

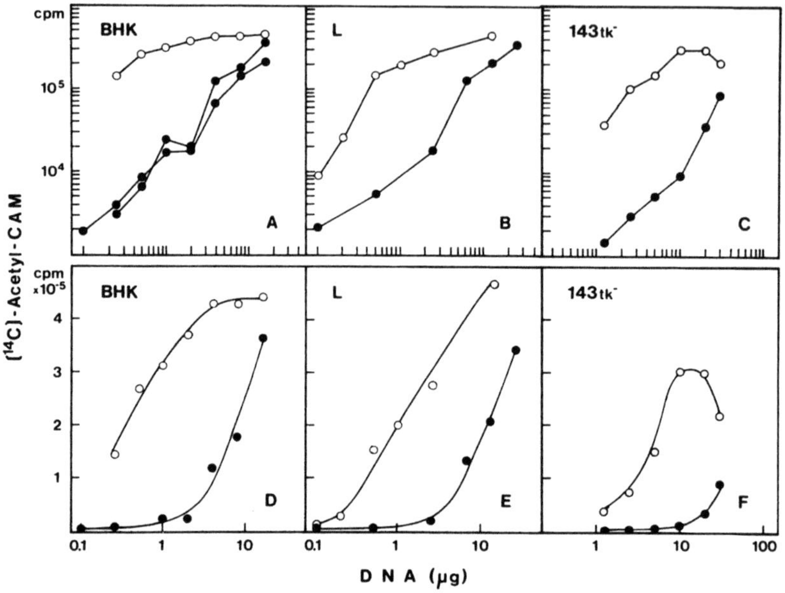

Fig. 5. Comparison of the C-*myc* and the SV40 early promoter activities in BHK, L, and 143tk⁻ cells by titration analysis. Increasing amounts of p*myc*CAT-P (*closed circles*) and pSV2CAT (*open circles*) were transfected as calcium phosphate precipitates into subconfluent BHK (A,D), L (B,E) and 143tk⁻ (C,E) cells. The vector plasmid pUC12 was used as carrier DNA to yield a total amount of 20 μg DNA per plate. The levels of CAT activity (cpm) determined using 350 μg (BHK, L) or 500 μg (143tk⁻) protein of the cell extract, were plotted as a function against the logarithm of transfected test DNA (μg). In the upper panels (A,B,C) a double logarithmic representation was used to demonstrate at low input of transfected DNA the approximately linear relationship between CAT activity and amount of DNA used. The half logarithmic plots shown in the lower panels (D,E,F) facilitate the determination of the amount of test DNA required for half maximal CAT activity

shown for the E1a-promoter of adenoviruses type 2 and 5 (Hearing and Shenk 1983, Hen et al. 1983).

To see whether sequences upstream of the c-*myc* promoters carry elements modulating the promoter activity, the CAT activity mediated by p*myc*CAT-P was compared to that of p*myc* CAT-R1, which carries additional 6.5 kb sequences upstream of the PvuII site. As shown in Fig. 6B, with higher amounts of transfected DNA (more than 0.52 pmol) a slightly lower, but similar activity was obtained with the long construct p*myc* CAT-R1. However, with lower amounts of transfected DNA (in the range of 0.026 to 0.52 pmol corresponding to 0.1-2.0 μg in the case of p*myc*CAT-P) the observed CAT activity diverges significantly. The long construct, p*myc*-CAT-R1, shows about a two- to eight-fold lower activity than the short p*myc* CAT-P plasmid, suggesting that a negative regulatory element might be located upstream of the PvuII site. Again, as with the short plasmid p*myc* CAT-PΔP2, removal of the second promoter in p*myc*CAT-R1ΔP2 decreased the CAT activity considerably. These results were confirmed by S1 analysis of RNA after transfection of p*myc* CAT plasmids into BHK cells revealing that the vast majority of the observed CAT activity was due to transcripts correctly initiated at the second of the two cap-sites also used in vivo (Lipp et al. 1987).

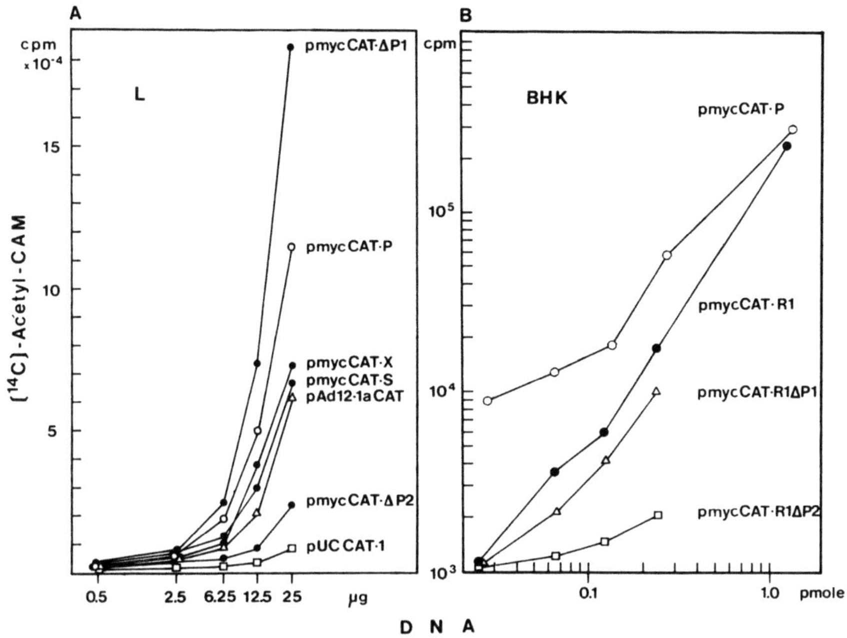

Fig. 6. Comparison of the CAT-activity mediated by the complete dual c-*myc* promoters and by various deletion constructs by titration analysis. Increasing equimolar amounts of the various CAT constructs, described in detail in Fig. 4, were transfected into subconfluent L (1) and BHK (B) cells. (A) The levels of CAT activity, determined using the total volume of the prepared cellular extract (150 µl), were plotted as a function against the logarithm of transfected test DNA. The plasmid pAd12-1aCAT (*open triangles*) was included in the titration analysis for comparison. (B) The CAT activity was determined from 200 µg protein of the extract. A double logarithmic representation was used to demonstrate the deviation in CAT activity at low input of transfected DNA between the different constructs. The amount of 0.263 pmole p*myc*CAT-P DNA (*open circles*) corresponds to 1 µg

Targets for Positive Regulation Within the c-myc Promoter. Evidence for two positively acting elements in the 862 bp PvuII-PvuII fragment has already been obtained from the experiments shown in the titration analysis in which deletions of the 248 bp upstream PvuII-SmaI fragment as well as of the second promoter reduced the activity of the *myc*- CAT constructs considerably. These observations were confirmed and extended in cotransfection experiments in which p*myc*CAT-R1, p*myc* CAT-P, p*myc* CAT-S and p*myc* CAT-PΔN as indicator plasmids were cotransfected with an excess of subcloned fragments carrying possible target sites for negatively or positively interacting cellular factors. Cotransfection of p*myc* CAT-R1 (Fig. 7A) and p*myc* CAT-P (Fig. 7B) with the 2.0 kb HindIII-PvuII upstream fragment did not alter the CAT activity as expected for p*myc*CAT-P sharing no sequences with the cotransfected fragment. However, for p*myc*CAT-R1 this result indicates that no negatively modulating element binds upstream of the PvuII site which can be titrated out leading to an activation of the promoter.

When, on the other hand, p*myc* CAT-R1 as an indicator plasmid was cotransfected with the 862 bp PvuII-PvuII fragment carrying both promoters, the 248 bp PvuII-SmaI fragment upstream of the first promoter, or the 614 bp SmaI-PvuII fragment with both promoters a significant

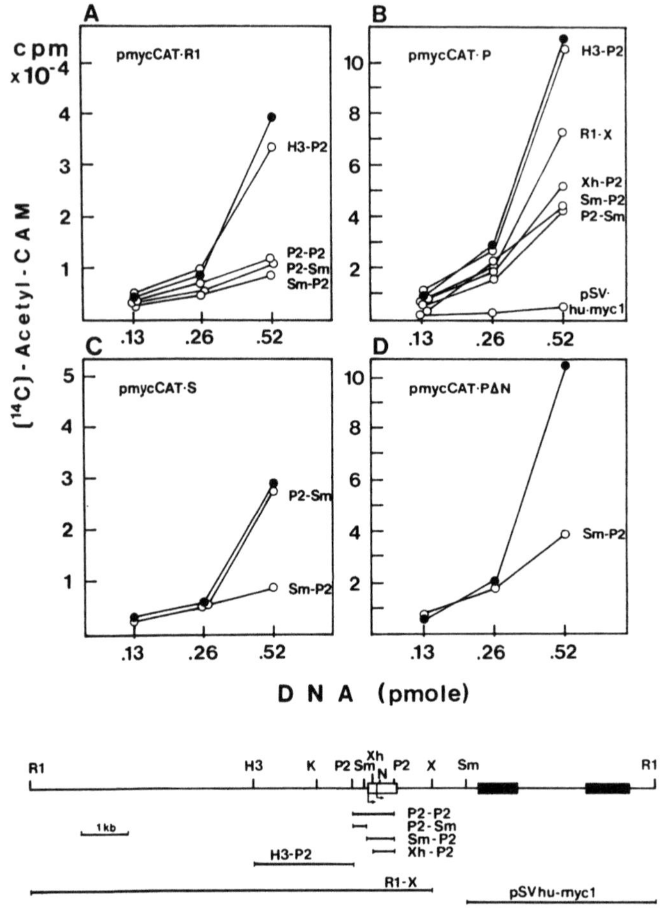

Fig. 7. Cotransfection of p*myc*CAT plasmids with competitor DNA's. Increasing amounts of p*myc*CAT indicator plasmids were transfected in the presence of constant amounts of competitor DNA's into subconfluent BHK cells. The CAT activity was determined 32-26 h after transfection using 400 μg protein of the cellular extract, and plotted as a function against the logarithm of transfected indicator DNA. The following indicator plasmids were used: (A) p*myc*CAT-R1; (B) p*myc*CAT-P (1 μg = 0.263 pmol); (C) p*myc*-CAT-S; (D) p*myc*CAT-PΔN; The titration kinetics of the indicator plasmids in the absence of competitor DNA and the presence of carrier DNA are indicated with closed circles. The competitor plasmids, carrying the fragments shown in the scheme below and next to the individual titration curves, were used in five to ten molar excess with respect to the highest concentration of the indicator plasmid. The CAT activities in the presence of competitor plasmids are symbolized with *open circles*. The same abbreviations for restriction enzymes are used as in Fig. 4

inhibitory effect was observed (Fig. 7A). As shown in Fig. 7B, both fragments upstream and downstream of the SmaI site, in addition the 450 bp XhoI-PvuII fragment carrying the second promoter and the 8.0 kb EcoRI-XbaI-fragment containing the complete first exon showed a similar inhibitory effect on p*myc* CAT-P. The specificity of this effect is demonstrated in Fig. 7C. Only cotransfection of the 614 bp SmaI-PvuII fragment, which shares sequences with p*myc*CAT-S, but none with the 248 bp PvuII-SmaI fragment decreased the activity of the indicator plasmid p*myc*CAT-S. The second positively acting element could be narrowed down to the 142 bp XhoI-NaeI fragment carrying the second promoter. As shown in Fig. 7D, deletion of the sequences of the first exon downstream from the NaeI site had no apparent effect on the activity of the construct compared to p*myc*CAT-P. Accordingly, the activity of the indicator plasmid p*myc* CAT-PΔN was significantly decreased by cotransfection with an excess of the subcloned SmaI-PvuII fragment. The most important result of these CAT experiments was the identification of sequence elements modulating the activity of the c-*myc* promoters. We were able to map two binding sites for positively regulating factors. The significance of these results is particularly interesting when related to the location of DNAaseI hypersensitive and protein binding sites in the upstream region of the c-*myc* gene (Siebenlist et al. 1984, Tuan and London 1984, Dyson and Rabbitts 1985). One positively acting element was mapped to the 248 bp PvuII-SmaI fragment located-101 to -349 bp upstream of the first capsite carrying within 36 bp four copies of a simple repeat element, TCCCCA, and two copies of the motif, CACCCTC, to which a DNAaseI hypersensitive site has been assigned. A similar direct repeat (CACCCTG) within the-100 box of the rabbit β-globin gene was found to be functionally important for efficient transcription. Interestingly, two copies of the motif TCCCCA are present in the 72 bp repeat next to the core sequence of the SV40 enhancer. This element, even though required for full promoter activity, is, however, not the most essential region. The promoter activity is more dramatically reduced if sequences downstream of the XhoI site were deleted. This second positive element could be assigned to the 142 bp XhoI-NaeI fragment (position -95 to +47) which carries the second promoter. Thus, we concluded that the dual promoter of the human c-*myc* gene represents a strong eukaryotic promoter regulated mainly by positively acting cellular transcription factors.

Comparison of the Translocated and Nontranslocated Allele. Interestingly, the p*myc* -CAT constructs carrying the first promoter only mediated very little CAT activity in monolayer cells, as well as in B-cells using transient expression assays, but this promoter is preferentially used in vivo in most Burkitt lymphoma cells. With the c-*myc* -CAT constructions from the nontranslocated c- *myc* allele of Raji cells we were unable to activate the first promoter neither in deletion constructs nor by cotransfection. One possibility is, that elements driving transcription from the first promoter are located upstream or downstream of the sequences present in the various CAT constructs. As a second possibility, the different promoter usage might reflect a peculiar chromatin structure in vivo, which is not formed when exogeneous DNA is transiently or stably introduced into the cells. Alternatively, the preferential usage of the first promoter might be a special feature of altered c-*myc* genes. We thought therefore it would be of particular importance to see whether the promoters of the c- *myc* genes derived from the normal or from the translocated chromosome of BL64 show functional differences in transient expression assays.

In order to test if the high expression of one c- *myc* allele in BL 64 was possibly caused by the somatic mutation in the first exon, the 862 bp PvuII fragments of both c- *myc* alleles was subcloned in front of the CAT gene as already described for the Raji derived constructs generating the clones p *myc*CAT64-P (with the mutation) and p*myc*CAT64-P_d (derived from the duplicated allele). The titration experiments shown in Fig. 8, revealed that the activity of both constructs followed a similar kinetic in baby hamster kidney cells (BHK) as well as in the human fibroblast line 143tk$^-$. To exclude the possibility that the different expression rates of the c- *myc* genes observed in the BL64 cell line, might reflect a cell type specificity, we transfected both p*myc*

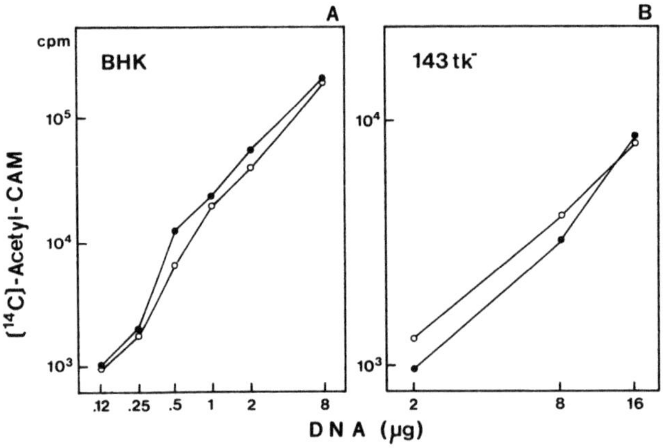

Fig. 8. Comparison of CAT activity mediated by the promoter-leader region of the c-*myc* gene isolated from both alleles of BL64 cells. Increasing amounts of the plasmids p*myc* CAT64-P (*filled circles*), containing the mutation, and p*myc*CAT64-P_d (*open circles*), derived from the duplicated allele (see Fig. 2) were transfected into subconfluent BHK (A) and 143tk⁻ (B) cells

Table 1. Comparison of CAT activity in BL64 cells mediated by the promoter-leader region of three different c-*myc* alleles

Transfected plasmid[a]	^{14}C-Acetyl-CAM (cpm)[b]	Percent CAT activity
pSV2CAT	3096 ± 440	100
p*myc*CAT-P	235 ± 62	8
p*myc*CAT64-P	247 ± 105	8
p*myc*CAT64-P_d	406 ± 74	13

[a]Suspension cells were transfected by the DEAE-Dextran procedure (Deans et al. 1984). 15 μg of plasmid DNA were incubated with 1 x 10^7 cells at 37°C for 45 min.

[b]Mean CAT activity and standard error determined from four separate transfections using 800 μg protein of the cellular extract

CAT64-P constructs into BL64 cells. The plasmid pSV2CAT was used as internal standard for transfection efficiency. As shown in Table 1, both c- *myc* promoters mediated a significant lower CAT acitivty compared to the SV40 promoter, but showed no differences with each other. From these results we concluded that the mutation in the first exon is not responsible for the high expression of this c-*myc* allele in the Burkitt cell. The generation of somatic mutations was proposed as a general mechanism leading to deregulation of the c-*myc* gene in cases in which the breakpoint does not disturb the first exon (Rabbitts et al. 1983, 1984, Denny et al. 1985). Since those somatic mutations are not found regularly, and we could show their nonefficacy in one case, we think that they are not necessary for the primary activation of the translocated c- *myc* gene, but they may contribute in some cases to progression of tumourigenicity.

Conclusions

Our detailed molecular analysis of the reciprocal breakpoints as well as of the c-*myc* loci of a Burkitt lymphoma carrying a variant t(2;8)

translocation provides new insights into the generation of chromosomal rearrangements, and in understanding of common features of Burkitt lymphomas. One of the most important findings is the detection of recognition sequences that apparently mediate site-specific recombination in the immediate vicinity of both breakpoints. Furthermore, we supplied direct experimental evidence that the c-*myc* gene can still be activated through the translocation despite a distance of at least 50 kb on the DNA level, and that somatic mutations, possibly inserted due to the influence of the immunoglobulin loci, are primarily not responsible for this activation.

Implication of c-*myc* in Burkitt Lymphoma and Cell Cycle

Our finding that the c-*myc* promoter carries target structures for positively acting transcription factors is in line with recently published data showing that the level of c-*myc* mRNA is invariant throughout the cell cycle and that the c-*myc* gene is transcribed at a normal rate in G_0-arrested chinese hamster lung fibroblasts but without detectable mature mRNA in the cytoplasm (Blanchard et al. 1985). Since after stimulation with growth factors the level of cytoplasmic c-*myc* mRNA increased without alteration of the transcription rate the authors proposed a model of post-transcriptional regulation of c-*myc* expression. The decrease of c-*myc* mRNA levels during retinoic acid-induced differentiation of F9 teratocarcinoma stem cells is also regulated postranscriptionally and associated with growth arrest (Dony et al. 1985, Dean et al. 1986). Interestingly, c-*myc* expression is also diminished during terminal differentiation of hematopoietic cells (Lachman and Skoultchi 1984, Gonda and Metcalf 1984, Coppola and Cole 1986). Thus, it seems that high levels of cytoplasmic c-*myc* mRNA are incompatible with the induction of differentiation.

These results, as well as the available data about BL's, together with our analysis of the variable t(2;8) translocation allow us to suggest a model explaning the initial steps responsible for the genesis of Burkitt lymphomas. One of the common features of all Burkitt lymphomas is that they are arrested in an early B-cell stage of B-cell differentiation. All carry surface IgM, or in rare cases surface IgM + D, or IgG (Fig. 9A). This, together with the second feature that most of the translocations have taken place in the nonfunctional or abberant rearranged allele of immunoglobulin genes, suggest that the translocations have to have occurred relatively early during B-cell development, simultanously with the functional immunoglobulin rearrangements. An essential role of the recombinase system in the generation of translocations is strongly implicated by our results. By the translocation the c-*myc* gene is thus located in the peculiar chromatin structure of immunoglobulin loci which could confer the accessibility to cis- and trans-acting cellular factors leading to its activation. The activated c-*myc* gene possibly prevents the B-cell from entering the G_0 stage of the cell cycle (Fig. 9B). Going through G_0 seems to be a condition for cells of a developing tissue to become susceptible to further differentiation. Thus, the block of B-cell differentiation caused by the translocation could result in the necessary growth advantage which might be a prerequisite for progression of oncogenicity due to the activation of other cellular and viral factors.

Acknowledgments. We thank G. Lenoir for providing us with the cell line BL64, I. Kruczek for gifts of pAd12-1aCAT and pSV2CAT, and S. Stelzig for synthesizing the oligonucleotide probe. The work was supported by the Deutsche Forschungsgemeinschaft (Li 374/1-1).

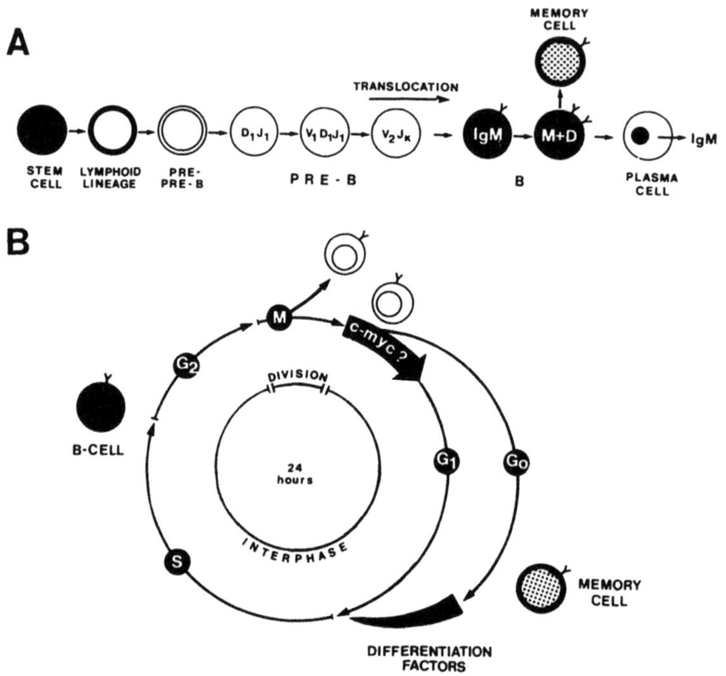

Fig. 9. The differentiation stage of Burkitt lymphoma within the B-cell lineage of the hematopoietic system (A), and model for a possible role of c-*myc* within the cell cycle (B, for explanation see text)

References

Adams JM, Gerondakis S, Webb E, Corcoran LM, Cory S (1983) Cellular myc oncogene is altered by chromosome translocation to an immunoglobulin locus in murine plasmacytomas and is rearranged similarly in human Burkitt lymphomas. Proc Natl Acad Sci USA 80:1982-1986

Armelin HA, Armelin MCS, Kelly K, Stewart T, Leder P, Cochran BH, Stiles CD (1984) Functional role for c-myc in mitogenic response to platelet-derived growth factor. Nature (London) 310:655-660

Ar-Rushdi A, Nishikura K, Erikson J, Watt R, Rovera G, Croce CM (1983) Differential expression of the translocated and the untranslocated c-myc oncogene in Burkitt lymphoma. Science 222:390-393

Battey J, Moulding C, Taub R, Murphy W, Stewart T, Potter H, Lenoir G, Leder P (1983) The human c-myc oncogene: Structural consequences of translocation into the IgH locus in Burkitt lymphoma. Cell 34:779-787

Bernard O, Cory S, Gerondakis S, Webb E, Adams JM (1983) Sequence of the murine and human cellular myc oncogenes and two modes of myc transcription resulting from chromosome translocation in B lymphoid tumours. EMBO J 2:2375-2383

Blanchard J-M, Piechaczyk M, Dani C, Chambard J-C, Franchi A, Pouyssegur J, Jeanteur P (1985) c-myc gene is transcribed at high rate in G_0 arrested fibroblasts and is post-transcriptionally regulated in response to growth factors. Nature (London) 317:443-445

Coppola JA, Cole MD (1986) Constitutive c-myc oncogene expression blocks erythroleukaemia cell differentiation but not commitment. Nature (London) 320:760-763

Croce CM, Thierfelder W, Erikson J, Nishikura K, Finan J, Lenoir GM, Nowell PC (1983) Transcriptional activation of an unrearranged and untranslocated c-myc oncogene by translocation of a C-lamda locus in Burkitt lymphoma cells. Proc Natl Acad Sci USA 80:6922-6926

Dani C, Blanchard JM, Piechaczyk M, El Sabouty S, Marty L, Jeanteur P (1984) Extreme instability of myc mRNA in normal and transformed human cells. Proc Natl Acad Sci USA 81:7046-7050

Davis M, Malcolm S, Rabbitts TH (1984) Chromosome translocation can occur on either side of the c-myc oncogene in Burkitt lymphoma cells. Nature (London) 308:286-288

Dean M, Levine RA, Campisi J (1986) c-myc regulation during retinoic acid-induced differentiation of F9 cells is posttranscriptional and associated with growth arrest. Mol Cell Biol 6:518-524

Deans RJ, Denis KA, Taylor A, Wall R (1984) Expression of an immunoglobulin heavy chain gene transfected into lymphocytes. Proc Natl Acad Sci USA 81:1292-1296

Denny CT, Hollis GF, Magrath IT, Kirsch IR (1985) Burkitt lymphoma cell line carrying a variant translocation creates new DNA at the breakpoint and violates the hierarchy of immunoglobulin gene rearrangement. Mol Cell Biol 5:3199-3207

Dony C, Kessel M, Gruss P (1985) Post-transcriptional contrl of myc and p53 expression during differentiation of the embryonal carcinoma cell line F9. Nature (London) 317:636-639

Duesberg PH, Bister K, Moscovici C (1980) Genetic structure of avian myeloblastosis virus, released from transformed myeloblasts as a defective virus particle. Proc Natl Acad Sci USA 77:5120-5124

Dyson PJ, Rabbitts TH (1985) Chromatin structure around the c-myc gene in Burkitt lymphomas with upstream and downstream translocation points. Proc Natl Acad Sci USA 82:1984-1988

Emanuel BS, Selden JR, Chaganti RSK, Jhanwar S, Nowell PC, Croce CM (1984) The 2p breakpoint of a 2;8 translocation in Burkitt lymphoma interrupts the v_{kappa} locus. Proc Natl Acad Sci USA 81:2444ß2446

Erikson J, Nishikura K, Ar-Rushdi A, Finan J, Emanuel B, Lenoir G, Nowell PC, Croce CM (1983) Translocation of an immunoglobulin kappa-locus to a region 3' of an unrearranged c-myc oncogene enhances c-myc transcription. Proc Natl Acad Sci USA 80:7581-7585

Gazin C, Dupont De Dinechin S, Hampe A, Masson J-M, Martin P, Stehelin D, Galibert F (1984) Nucleotide sequence of the human c-myc locus: provocative open reading frame within the first exon. EMBO J 3:383-387

Gonda TJ, Metcalf D (1984) Expression of myb, myc and fos protooncogenes during the differentiation of a murine myeloid leukemia. Nature (London) 310:249-251

Gorman CM, Moffat LF, Howard BH (1982) Recombinant genomes which express chloramphenicol acetyltransferase in mammalian cells. Mol Cell Biol 2:1044-1051

Hameister H, Adolph S (1986) Breakpoint determination of variant t(2;8) translocations in Burkitt lymphoma cell lines. Hum Genet (in press)

Hamlyn PH, Rabbitts TH (1983) Translocation joins c-myc and immunoglobulin gamma-1 genes in a Burkitt lymphoma revealing a third exon in the c-myc oncogene. Nature (London) 304:135-139

Hann SR, Eisenman RN (1984) Proteins encoded by the human c-myc oncogene: Differential expression in neoplastic cells. Mol Cell Biol 4:2486-2497

Hayday AC, Gillies SD, Saito H, Wood C, Wiman K, Hayward WS, Tonegawa S (1984) Activation of a translocated human c-myc gene by an enhancer in the immunoglobulin heavy-chain locus. Nature (London) 307:334-340

Hearing P, Shenk T (1983) The Adenovirus type 5 E1a transcriptional control region contains a duplicated enhancer element. Cell 33:695-703

Hen R, Borelli E, Sassone-Corsi P, Chambon P (1983) An enhancer element is located 340 base pairs upstream from the adenovirus-2 E1a capsite. Nucleic Acids Res 11: 8747-8758

Hood L, Kronenberg M, Hunkapiller T (1985) T cell antigen receptor and the immunoglobulin supergene family. Cell 40:225-229

Hope TJ, Aguilera RJ, Minie ME, Sakano H (1986) Endonucleolytic activity that cleaves immunoglobulin recombination sequences. Science 231:1141-1145

Kelly K, Cochran BH, Stiles CD, Leder P (1983) Cell-specific regulation of the c-myc gene by lymphocyte mitogens and platelet-derived growth factor. Cell 35:603-610

Klein G (1983) Specific chromosomal translocations and the genesis of B-cell-derived tumors in mice and men. Cell 32:311-315

Klein G, Klein E (1985) Evolution of tumours and the impact of molecular oncology. Nature (London) 315:190-195

Kruczek I, Doerfler W (1983) Expression of the chloramphenicol acetyltransferase gene in mammalian cells under the control of adenovirus type 12 promoters: Effect of promoter methylation on gene expression. Proc Natl Acad Sci USA 80:7586-7590

Lachman HM, Skoultchi AI (1984) Expression of c-myc changes during differentiation of mouse erythroleukaemia cells. Nature (London) 310:592-594
Lipp M, Hartl P (1986) Possible role of immunoglobulin recombination sequences in the genesis of variant t(2;8) translocations of Burkitt lymphoma. Curr Top Microbiol Immunol 132:162-168
Lipp M, Schilling R, Wiest S, Laux G, Bornkamm GW (1987) Target sequences for cis-acting regulation within the dual promoter of the human c-myc gene. Mol Cell Biol (in press)
McCormack JE, Pepe VH, Kent RB, Dean M, Marshak-Rothstein A, Sonenshein GE (1984) Specific regulation of c-myc oncogene expression in a murine B-cell lymphoma. Proc Natl Acad Sci USA 81:5546-5550
Nishikura K, Ar-Rushidi A, Erikson J, Watt R, Rovers G, Croce CM (1983) Differential expression of the normal and of the translocated human c-myc oncogenes in B cells. Proc Natl Acad Sci USA 80:4822-4826
Persson H, Leder P (1984) Nuclear localization and DNA binding properties of a protein expressed by human c-myc oncogene. Science 225:718-721
Persson H, Hennighausen L, Taub R, DeGrado W, Leder P (1984) Antibodies to human c-myc oncogene product: evidence of an evolutionarily conserved protein induced during cell proliferation. Science 225:687-693
Piechaczyk M, Yang J-Q, Blanchard J-M, Jeanteur P, Marcu KB (1985) Postranscriptional mechanisms are responsible for accumulation of trancated c-myc RNAs in murine plasma cell tumors. Cell 42:589-597
Rabbitts TH, Hamlyn PH, Baer R (1983) Altered nucleotide sequences of a translocated c-myc gene in Burkitt lymphoma. Nature (London) 306:760-765
Rabbits TH, Forster A, Hamlyn P, Baer P (1984) Effect of somatic mutation within translocated c-myc genes in Burkitt's lymphoma. Nature (London) 309:592-597
Rappold GA, Hameister H, Cremer T, Adolph S, Henglein B, Freese U-K, Lenoir GM, Bornkamm GW (1984) c-myc and immunoglobulin kappa light chain constant genes are on the $8q^{+}$ chromosome of three Burkitt lymphoma lines with t(2;8) translocation. EMBO J 3:2951-2955
Seidman JG, Max EE, Leder P (1979) A kappa-immunoglobulin gene is formed by site-specific recombination without further somatic mutation. Nature (London) 280: 370-375
Siebenlist U, Henninghausen L, Battey J, Leder P (1984) Chromatin structure and protein binding in the putative regulatory region of the c-myc gene in Burkitt lymphoma. Cell 37:381-391
Taub R, Kelly K, Battey J, Latt S, Lenoir GM, Tantravahi U, Tu Z, Leder P (1983) A novel alteration in the structure of an activated c-myc gene in a variant t (2;8) Burkitt lymphoma. Cell 37:511-520
Taub R, Moulding C, Battey J, Murphy W, Vasicek T, Lenoir GM, Leder P (1984) Activation and somatic mutation of the translocated c-myc gene in Burkitt lymphoma cells. Cell 36:339-348
Tonegawa S (1983) Somatic generation of antibody diversity. Nature (London) 302: 575-581
Tsujimoto Y, Jaffe E, Cossman J, Gorham J, Nowell PC, Corce CM (1985) Clustering of breakpoints on chromosome 11 in human B-cell neoplasms with the t(11;14) chromosome translocation. Nature (London) 315:340-343
Tuan D, London Im (1984) Mapping of DNase I-hypersensitive sites in the upstream DNA of human embryonic epsilon-globin gene in K562 leukemia cells. Proc Natl
Watt R, Nishikura K, Sorrentino J, Ar-Rushdi A, Croce CM, Rovera G (1983) The structure and nucleotide sequence of the 5' end of the human c-myc oncogene. Proc Natl Acad Sci USA 80:6307-6311

Biochemical Properties of Oncogene-Coded Proteins

K. Moelling[1], U. Lorenz[1], P. Beimling[1], B. Heimann[1], G. Thomas[2],
H. Bading[1], and T. Sander[1]

Introduction

The analysis of viral oncogenes and their gene products has allowed extensive studies on molecular mechanisms of malignant transformation in vitro. Understanding of tumorigenesis of human cells has profited extensively from knowledge of oncogenes, since similar genes are expressed in many human tumor cells. About two dozen different viral oncogenes have been detected by the analysis of retroviruses so far (Bishop 1985). These genes can be classified into three functional groups. The largest group comprises the *src*- gene family which characteristically codes for tyrosine-specific protein kinases (Sefton 1985). This group of oncogenes exhibits relationship to growth hormones or growth hormone receptors and differs from their normal cellular counterparts by modifications which lead to constitutive tyrosine kinase expression and loss of responsiveness to mitogenic signals (Hunter 1984). The mode of signal transmission from the membrane-associated tyrosine kinase through putative second messengers to the nucleus which results in proliferation, is largely unknown. Phosphatidyl-inositol and protein kinase C may be indirectly involved (Moelling et al. 1984, Sugano and Hanafusa 1985). The second type of oncogenes is represented by only very few membres, e.g., *mil*, which is associated with a serine/threonine specific protein kinase (Moelling et al. 1984). It may regulate protein expression through phosphorylation of the ribosomal subunit S6. The third group of oncogene proteins, e.g., *myc* and *myb*, is predominantly located in the nucleus and binds to DNA in vitro (Donner et al. 1982, Franchini et al. 1983, Goh et al. 1985, Moelling et al. 1985). They presumably play a role in regulation of gene expression. The recently discovered *tat* genes of HTLV viruses are putative oncogenes as well. The *tat* gene product of HTLV I (tat_I) is expected to interact with the viral LTR and thereby regulate gene expression and also affect cellular gene expression (Sodroski et al. 1984).

Results

In order to analyze oncogene protein products, antibodies are required for their identification. Since the oncogene proteins exhibit rather low immunogenicity in the infected animal, antibodies were raised in two ways: first, cloned DNA of oncogenes was inserted into expression vectors for protein expression in bacteria, secondly, oligopeptides specific for oncogenes were synthesized in vitro and after coupling to

[1]Max-Planck-Institut für Molekulare Genetik, Abt. Schuster, Ihnestrasse 73, D-1000 Berlin 33, FRG

[2]Friedrich-Miescher-Institut, Postfach 2543, CH-4002 Basel, Schweiz

37. Colloquium Mosbach 1986
Cell Cycle and Oncogenes

carrier proteins used as immunogens. The oncogene-specific bacterial proteins or synthetic peptides served furthermore for purification of the specific immunoglobulins (IgG's) from the sera. These IgG's were then coupled to immunobeads and served as immunoaffinity columns for the identification and isolation of oncogenecoded proteins from either virus-transformed cells or human tumor cell lines. Furthermore, the antigens were used for the identification and isolation of monoclonal antibodies against *myc* and *myb* (manuscript in preparation).

The Oncogene Proteins *myc*, *myb* and tat_I

The viral (v-) *myc*, human cellular *myc* (hu-c-*myc*) and v-*myb* oncogenes were expressed in bacteria using the thermoinducible pPLc24 expression vector system (Bunte et al. 1984, Beimling et al. 1985, Moelling et al. 1985). tat_I expression using a β-galactosidase expression vector did not give rise to high titered antibody. A carboxy-terminal synthetic peptide was used more successfully. The oncogene-specific IgG's were coupled to protein A-Sepharose which served as immunoaffinity column. Cellular lysates from metabolically labeld ^{35}S-methionine-labeled cells were applied to these columns which retained the oncogene proteins. Treatment with citric-acid buffer allowed the removal of the proteins from the columns. The buffer was neutralized immediately after elution of the proteins to minimize their damage by pH shock. Efficiency of purification by this approach was approximately 3000-fold. The v-*myc* protein, $p110^{gag-myc}$, was purified from MC29-Q8 quail fibroblastic cell line, the $p64^{hu-c-myc}$ from Colo 320, $p48^{v-myb}$ from BM-2, an AMV-transformed bone marrow cell line, and $p75^{hu-c-myb}$ from MOLT4, a human hemopoietic cell line (Franchini et al. 1983). The tat_I protein, p42 tat_I, was isolated from C8166 cells a nonproducer cell line expressing tat_I protein at high levels (Lee et al. 1984). The purified proteins are shown in Fig. 1A. The amounts of proteins shown reflect their relative cellular concentrations.

Cell fraction analysis indicated that about 50 to 60% of the v/c-*myc*, v/c-*myb* and tat_I proteins are located in the nuclei of transformed cells (Donner et al. 1982, Beimling et al. 1985, Goh et al. 1985, Moelling et al. 1985). This suggested a role at the level of gene regulation for these proteins. To test this hypothesis, the purified proteins were anlayzed for their abilities to bind to DNA in vitro. Calf-thymus DNA sheared to about 2 kb in size was labeled in vitro with bacterial DNA polymerase I (Klenow fragment) or end-labeled with T4 polymerase using ^{32}P-labeled nucleotides (Maniatis et al. 1982). Alternatively normal chicken cell DNA from ^{3}H-thymidine metabolically labeled cells was extracted, sheared and used (Donner et al. 1982). DNA-binding was tested by filter-binding assay using radioactively labeled DNA and purified ^{35}S-methionine-labeled proteins in filter-binding assays. Increasing amounts of proteins retained increasing amounts of DNA on the filter. The radioactivity of this DNA was determined and plotted versus protein concentrations. The results for v- *myc* hu-c-*myc*, v-*myb* and tat_I are shown in Fig. 1B. The hu-c-*myb* protein was not available in sufficient amounts to perform this kind of assay.

On the assumption of a regulation of gene expression through the DNA-binding proteins, one would predict specific DNA sequences to be preferentially detected. According to a recently suggested model, the *myc* protein was proposed to interact with the viral promoter/enhancer sequence, the LTR (Robertson 1983). Furthermore, the tat_I protein has been shown to either directly or indirectly interact with specific sequences in the LTR of HTLV I (Sodroski et al. 1984). To test the ex-

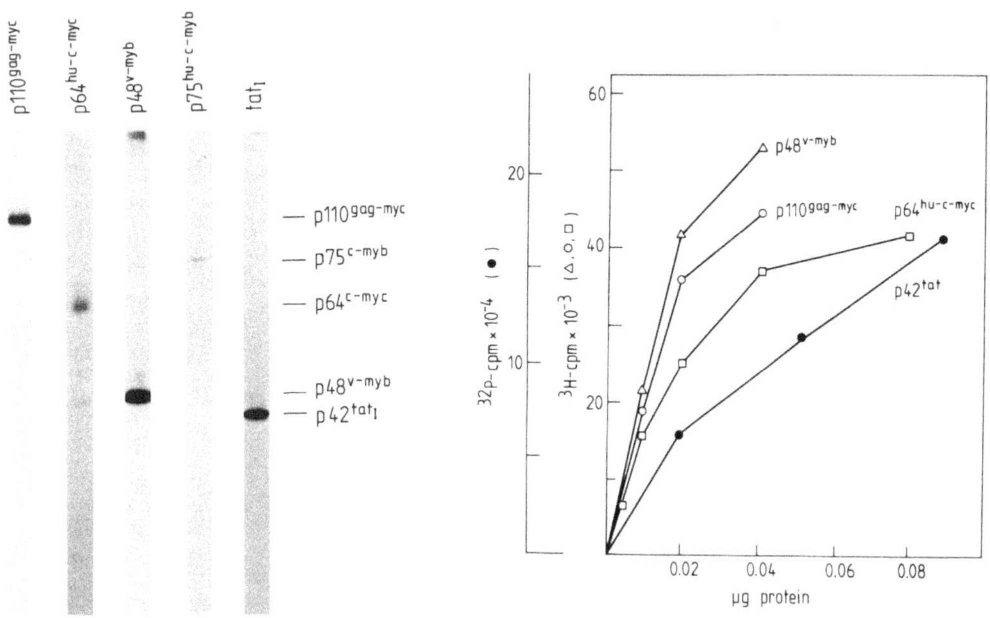

Fig. 1(A) Oncogene proteins purified by immunoaffinity chromatography using specific IgG's coupled to protein A-Sepharose. Proteins were purified from ^{35}S-methionine metabolically labeled cell lines such as MC29 quail fibroblasts (Q8), Colo 320, BM-2, MOLT4 and C8166 (from left to right). The amounts of proteins shown correspond to 5 x 10^6 cells. The purified proteins were analyzed by SDS gel electrophoresis and autoradiography (3 to 5 days). (B) DNA-binding of purified proteins. The purified ^{35}S-methionine-labeled proteins shown in (A) were analyzed for their abilities to bind to DNA in vitro by filter-binding assay. ^{3}H-thymidine-labeled normal cell DNA or calf thymus DNA labeled with ^{32}P nucleotides in vitro were used as indicated. The amount of protein was determined from the specific activities of the corresponding cellular lysates. The radioactivity retained on the filter was determined by liquid scintillation counting

pected specific interaction, DNA fragments from DNA clones were prepared and labeled in vitro. The *myc* and tat_I protein were allowed to interact with the DNA. Subsequently the DNA fragments retained on the filters were eluted and then analyzed on agarose gels.

As is shown in Fig. 2, the pMC29 DNA clone (Lautenberger et al. 1981) was restricted by endonucleases which gives rise to 6 prominent bands ranging in size from 0.55 to 4.4 kb. All these fragments were retained on the filter by the v-*myc* protein and identified after elution from the filter by agarose gel electrophoresis. Presence of competing unlabeled double-stranded DNA at two concentrations results in the preferential binding of a 2.4 kb fragment by the v-*myc* protein. The 4.4 kb fragment containing essentially sequences of the cloning vector pBR325 is also retained to some extent. The 2.4 kb fragment is derived from flanking cellular sequences upstream of the MC29 viral DNA integration site (Lautenberger et al. 1981). The v-*myc* protein does not preferentially recognize the viral LTR which is located on the 3.2 kb fragment - a result which contradicts the suggested model mentioned above (Robertson 1983). Interaction of tat_I with DNA fragments is being analyzed with a DNA construct containing the LTR of HTLV I.

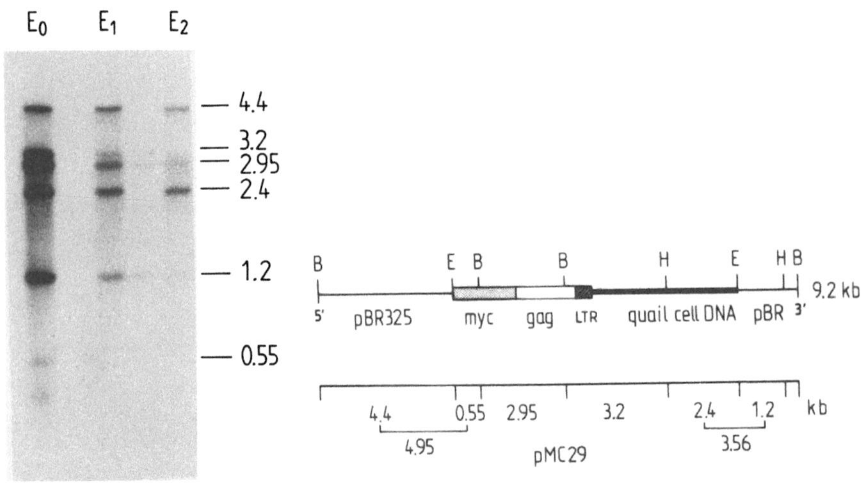

Fig. 2. Binding of v-*myc* protein to DNA fragments. pMC29 DNA (30 ng) restricted by *Bam*HI, *Hin*dIII and *Eco*RI ([B/H/E]) and radioactively labeld in vitro was incubated with purified p110$^{gag-myc}$ protein (30 ng) and analyzed by filter-binding assay. DNA retained on the filter was reeluted and analyzed by agarose gel electrophoresis and subsequent autoradiography (3 h). E_0, E_1 and E_2 indicate the eluted material from reactions which contained zero, 0.3, and 0.6 µg of competing calf thymus DNA during the incubation period. The numbers indicate the size of the DNA fragments in kilobase (kb). The restriction map of the pMC29-DNA clone with *Bam*HI (*B*), *Eco*RI (*E*), *Hin*dIII (*H*) is shown

The Oncogene Protein *mil*

The MH2 virus is one of the few retroviruses which codes for two oncogenes, *mil* and *myc* (for review see Moelling 1985). The *mil* oncogene is highly related to the oncogene *raf* (Jansen et al. 1984). In MH2 virus the *mil* is expressed as a viral fusion protein, p100$^{gag-mil}$. This protein was purified by immunoaffinity column chromatography using monoclonal antibodies against the N-terminus of the *gag* portion. In order not to destroy enzymatic activity of the protein, it was not eluted from the immunoaffinity column, but remained bound to the Sepharose beads through the immunoglobulin molecules. This immunobead-coupled *mil* protein was shown to exhibit protein kinase activity which differed, however, from the protein kinases of the *src* family in that it phosphorylates serine/threonine but not tyrosine (Moelling et al. 1984). Recently, mutant studies have indicated that the protein kinase activity disappears if a lesion is introduced into the *mil* gene, supporting that this kinase is virus-coded (unpublished observation).

It is still an open question, which target molecules are phosphorylated by *src*-related protein kinases or the *mil* kinase. We were able to show that neither *mil* nor the *src*- related oncogene protein *fps* are directly responsible for lipid phosphorylation (Moelling et al. 1984). Therefore, changes in lipid metabolism such as phosphorylation of phosphatidylinositol into di- and tri-phosphates is an indirect effect of *src*- and *mil* protein kinases.

In rapidly dividing cells transformed by retroviruses the ribosomal subunit S6 has been consistently observed to be highly phosphorylated, an

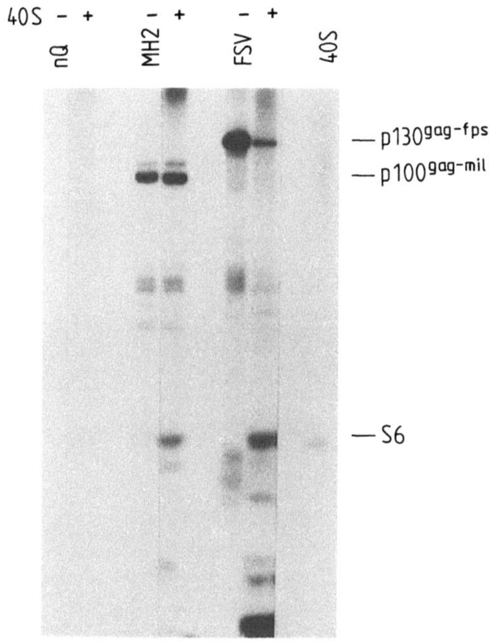

Fig. 3. Effect of S6 phosphorylation. Purified p100gag-mil and p130gag-fps from MH2 quail fibroblasts and Fujinami sarcoma virus-transformed rat cell line, respectively, were tested in protein kinase assays in the absende (-) and presence (+) of 40S ribosomal subunit S6. Normal quail fibroblasts (nQ) were processed identically as p100gag-mil and used as control. Autophosphorylation of 40S is shown for comparison. The protein kinase reaction products were analyzed by gel electrophoresis and autoradiography. Exposure time 1 h

effect which can also be induced by serum stimulation of quiescent cells, growth factors or tumor promoters. S6 phosphorylation may be linked to initiation of protein synthesis and thereby control cell proliferation (Blenis and Erikson 1984).

To test the possible effect of *mil* protein kinase on S6 phosphorylation, 40S ribosomal subunits were used as substrates for the *mil* and *fps* protein kinases. As is shown in Fig. 3, the 30K S6 protein from the 40S ribosomal subunit is phosphorylated in vitro. Autophosphorylaton of *mil* and *fps* occurs as well. Phosphoamino acid analysis of the phosphorylated S6 proteins indicated serine/threonine phosphorylation in the case of *mil*, and tyrosine in the case of *fps* protein kinases. Autophosphorylation of *mil* and *fps* exhibited the analogous specificities (data not shown).

Comparison of the optimal enzyme conditions of the S6 protein kinase in vitro with *mil* protein kinase indicates differences, *mil* prefers Mn whereas S6 protein kinase prefers Mg as divalent cation. Further experiments are under way to analyze the relationship between the two protein kinases.

Discussion

The results presented here demonstrate two novel observations, the preferential binding of the *myc* protein to a cellular DNA sequence and the ability of the *mil* protein kinase to phosphorylate the ribosomal protein S6 in vitro.

Previously the *myc* gene product has been demonstrated not to selectively bind to DNA fragments (Watt et al. 1985). The model on the role

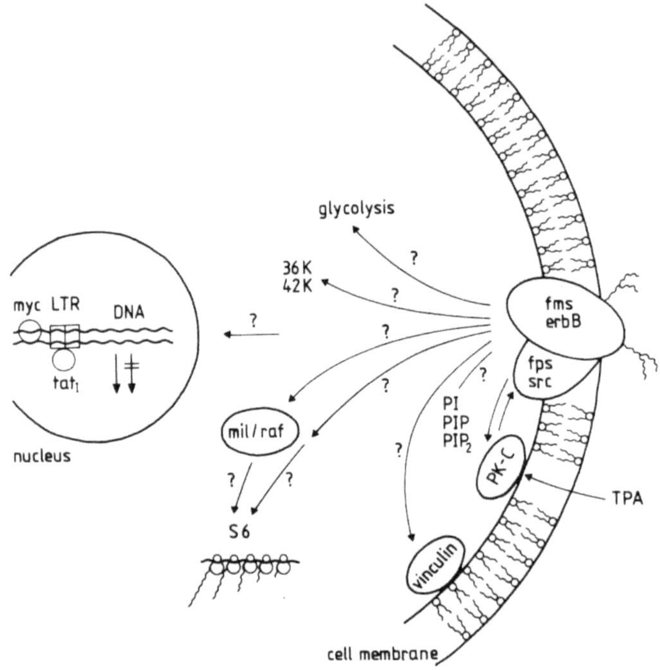

Fig. 4. Model on mode of action of various oncogene proteins. Representatives of the three types of oncogene protein are shown schematically. Associated with the membrane are the tyrosine-specific protein kinases some of which resemble growth-factor receptors (*erb*B, *fms*, *src*, *fps*). The oncogene proteins *mil/raf* which are located in the cytoplasm are serine/threonine specific protein kinases. The effect of the protein kinases is uncertain. Potential mechanisms involve glycolysis, cellular target proteins (42K, 36K), phosphatidyl-inositol (PI)-phosphorylation (PIP, PIP_2), vinculin, protein kinase C (PK-C) which is also affected by tumor promoters (TPA). Phosphorylation of ribosomal protein S6 may be affected directly or indirectly. The nuclear oncogene proteins such as *myc*, *myb*, tat_I are considered as gene regulatory proteins. The *myc* protein does not bind to LTR-like sequences but upstream cellular sequences. The effect of tat_I, however, is thought to involve its own LTR and other promoter/enhancers

of the *myc* protein suggested an interaction with the viral LTR. According to the results presented here, this model may not be correct. The result shown in Fig. 2 points out to another possible mechanism which involves cellular sequences upstream of the LTR and the integration site of the MC29 DNA provirus into the cellular genome. Recently it has been shown that such upstream sequences harbor several hypersensitive sites (Siebenlist et al. 1984), which are involved in gene expression. Further experiments are required to show details of the type of interaction and biological effects of binding of the *myc* protein to such sequences. The expected effect of tat_I on cellular DNA sequences presumably involves the LTR promoter/enhancer type DNA. This has been demonstrated indirectly so far (Sodroski et al. 1984). Experiments to prove a direct binding of tat_I to LTR are in progress.

The effect of the protein kinases on intracellular events are still largely unknown. Recently, the effect of tyrosine protein kinases on phosphorylation of the 40S ribosomal protein S6 has been shown (Blenis and Erikson 1985). This effect must, however, be of indirect nature, since phosphorylation of S6 occurs in serine not in tyrosine as would be expected for a direct tyrosine kinase effect. The *mil* protein kinase which is serine/threonine specific, may be a candidate for S6 phosphorylation. In vitro such a phosphorylation can be achieved as is demonstrated here. Oncogene-coded protein kinases are, however, promiscuous in vitro. Therefore further experiments are required to demonstrate whether *mil* is related to the S6 protein kinase.

Materials and Methods

Purification of Oncogene Proteins. 5 x 10^7 cells of MC29-Q8 fibroblasts, Colo 320, BM-2 bone marrow cells transformed by avian myeloblastosis virus (AMV), MOLT4 (Franchini et al. 1983), and the tat_I producing human lymphocyte cell line C8166 (Lee et al. 1984) were labeled metabolically with ^{35}S-methionine (500 µCi/ml) for 2 h and processed for protein purification by immunoaffinity chromatography. MH2-transformed quail fibroblasts and Fujinami sarcoma virus-transformed rat cells were processed without radioactive labeling. The immunoaffinity columns consisted of IgG specific for the N-terminal portion of *gag* for all *gag-onc* fusion proteins as described (Donner et al. 1982, Moelling et al. 1984). In the case of *myc* and *myb*, *myc-* and *myb-* specific IgG's were used (Moelling et al. 1985), and in the case of *tat*, peptide-specific IgG was isolated. The proteins were isolated from the immunoaffinity columns by citric acid buffer, pH 3.5, which was immediately neutralized after elution (Moelling et al. 1985). The protein kinases *mil* and *fps* do not allow this treatment, therefore the proteins remained bound to the immunoaffinity columns and were used for protein kinase assays as bead-coupled enzymes (Moelling et al. 1984).

Filter Binding Assays. Calf thymus DNA was labeled in vitro using Klenow fragment and ^{32}P-labeled nucleotides. The DNA was sheared to 2 kb and labeled to a specific activity of 2.5 x 10^7 cpm/µg. Ten ng of DNA corresponding to 250,000 cpm total were used per assay. Normal chicken cell DNA metabolically labeled with ^{3}H-thymidine was sheared to 20 kb in size. The specific activity of the DNA was about 200,000 cpm/µg. The amount of DNA used per assay was 250 ng corresponding to 50,000 cpm. DNA-filter binding assays have been described (Scheidereit et al. 1983). The amounts of proteins used were determined from the specific activity of the cellular lysate and the amount of radioactivity of the purified proteins.

Analysis of DNA Fragments. pMC29 DNA was restricted by *Bam*HI, *Hin*dIII, *Eco*-RI and end-labeled with the Klenow fragment of *E. coli* DNA polymerase I and [α-^{32}P]dCTP as described by Maniatis et al. (1982). Specific activity of the DNA was 1-5 x 10^6 cpm/µg. The DNA fragments (30 ng) and purified $p110^{gag-myc}$ (30 ng) were incubated in 100 µl binding buffer (10 mM Tris-HCl, pH 7.5, 0.1 mM EDTA, 1 mM $MgOAc_2$, 0.1 mM dithiothreitol, 0.1 µg/ml bovine serum albumin and 60 mM KCl) for 10 min at 25°C. For competition experiments sheared native calf thymus DNA was added together with the labeled DNA fragments at varying concentrations.

The samples were filtered at room temperature through nitrocellulose filters as described (Scheidereit et al. 1983). Using buffer containing 0.1% sodium dodecyl sulfate the DNA fragments were eluted from the filters (3 hrs, 37°C) precipitated and electrophoresed in agarose (1%) gels. The gels were subsequently dried and processed for autoradiography.

Protein Kinase Assays. The immunobead-coupled *mil* and *fps* protein kinases were analyzed for protein kinase as described (Moelling et al. 1984).

10 µg of 40S ribosomal subunits were added to the protein kinase reactions. The reaction products were subsequently applied to 10% SDS-polyacrylamide gels and analyzed by autoradiography.

References

Beimling P, Benter T, Sander T, Moelling K (1985) Isolation and characterization of the human cellular myc gene product. Biochemistry 24:6349-6355

Bishop JM (1985) Trends in oncogenes. Trends Genet Sept:245-249

Blenis J, Erikson RL (1984) Phosphorylation of the ribosomal protein S6 is elevated in cells transformed by a variety of tumor viruses. J Virol 50:966-969

Blenis J, Erikson RL (1985) Regulation of ribosomal protein S6 kinase activity by the Rous sarcoma virus transforming protein, serum, or phorbol ester. Proc Natl Acad Sci USA 82:7621-7625

Bunte T, Donner P, Pfaff E, Reis B, Greiser-Wilke I, Schaller H, Moelling K (1984) Inhibition of DNA binding of purified $p55^{v-myc}$ in vitro by antibodies against bacterially expressed myc protein and a synthetic peptide. EMBO J 3:1919-1924

Donner P, Greiser-Wilke I, Moelling K (1982) Nuclear localization and DNA-binding of the transforming gene product of avian myelocytomatosis virus. Nature (London) 296:262-266

Franchini G, Wong-Staal F, Baluda MA, Lengel C, Tronick SR (1983) Structural organization and expression of human DNA sequences related to the transforming gene of AMV. Proc Natl Acad Sci USA 80:7385-7389

Goh WC, Sodroski J, Rosen C (1985) Subcellular localization of the product of the long open reading frame of human T-cell leukemia virus type I. Science 227:1277-1231

Hunter T (1984) The epidermal growth factor receptor gene and its product. Nature (London) 311:414-416

Jansen H, Lurz R, Bister K, Bonner TI, Mark GE, Rapp UR (1984) Homologous cell-derived oncogenes in avian carcinoma virus MH_2 and murine sarcoma virus 3611. Nature (London) 307:281-284

Lautenberger JA, Schulz RA, Gavon CF, Tsichlis PN, Papas TS (1981) Molecular cloning of MC29 transforming sequences. Proc Natl Acad Sci USA 78:1518-1522

Lee TH, Coligan JE, Sodroski JG, Haseltine WA, Wong-Staal F, Gallo RC, Essex M (1984) Antigens encoded by the 3' terminal region of human T-cell leukemia virus: evidence for a functional gene. Science 226:57-61

Maniatis T, Fritsch EF, Sambrook J (1982) Molecular cloning. Cold Spring Harbor Lab Press, Cold Spring Harbor New York

Moelling K (1985) Fusion proteins in retroviral transformation. In: Klein G, Weinhouse S (eds) Advances in cancer research, vol 43. Academic Press, London New York, pp 205-239

Moelling K, Heimann B, Beimling P, Rapp U, Sander T (1984) Serine- and threonine-specific protein kinase activities of purified gagmil and gag-raf proteins. Nature (London) 312:558-561

Moelling K, Pfaff E, Beug H, Beimling P, Bunte T, Schaller HE, Graf T (1985) DNA-binding activity is associated with purified myb proteins from AMV and E26 viruses and is temperature-sensitive for E26 ts mutants. Cell 40:983-990

Robertson M (1983) Paradox and paradigm: the message and meaning of myc. Nature (London) 306:733-736

Scheidereit C, Geisse S, Westphal HM, Beato M (1983) The glucocorticoid receptor binds defined nucleotide sequences near the promoter of mouse mammary tumor virus. Nature (London) 304:749-752

Sefton BM (1985) Oncogenes encoding protein kinases. Trends Genet Nov:306-308

Siebenlist U, Henninghausen L, Batty J, Leder P (1984) Chromatin structure and protein binding in the putative regulatory region of the c-myc gene in Burkitt Lymphoma. Cell 34:381-391

Sodroski JG, Rosen CA, Haseltine WA (1984) Trans-acting transcripitonal activation of the long terminal repeat of human T-lymphotropic viruses in infected cells. Science 225:381-385

Sugano S, Hanafusa H (1985) Phosphatidylinositol kinase activity in virus-transformed an nontransformed cells. Mol Cell Biol 5:2399-2404
Watt RA, Shatzman AR, Rosenberg M (1985) Expression and characterization of the human c-myc DNA binding protein. Mol Cell Biol 5:448-456

Suppressors of *RAS* Function in *Saccharomyces cerevisiae*

K. Tatchell, J. F. Cannon, L. C. Robinson, and R. B. Wilson[1]

Introduction

The yeast *Saccharomyces cerevisiae* contains at least three genes which share homology with members of the *ras* oncogene family (6,7,12,23). *YPT*, on chromosome VI, shares about 40% homology with mammalian *ras* in the first 160 amino acids. *RAS1* and *RAS2* on chromosome XV and XIV respectively, are over 60% homologous with *ras* in the same region. Genetic analysis of these genes reveals that *YPT* and *RAS1* and *RAS2* constitute essential functions. Yeast cells lacking functional *RAS1* and *RAS2* genes are inviable (17,28); *YPT* fails to complement the mutations in *RAS1* and *RAS2* and therefore encodes a separate function (Gallwitz et al. Chap. II 8, this Vol.).

Members of the *ras* gene family in mammals have been implicated in a wide variety of cancers. Originally identified as the transforming gene in Harvey and Kirsten sarcoma viruses, *ras* proto-oncogenes were subsequently found in normal cells (reviewed in (9)). The viral genes differ from their normal cellular homologs by the lack of introns and by missense mutations at a limited number of sites. Although the normal cellular genes can transform immortalized cell lines, *ras* genes with missense mutations corresponding to the ones found in the viral genes have a much greater transformation potential. *Ras* gene products (p21) from transforming genes cause transient transformation when injected directly into tissue culture cells. p21 molecules with transforming ability have a decrease in intrinsic GTPase activity. The affinity of p21 for GTP and GDP, however, is not altered by the transforming mutations (reviewed in (13)).

The yeast *RAS* proteins share many of the biochemical properties of their mammalian counterparts. They can be immunoprecipitated by antibodies against p21, they bind GTP and GDP (26,30) and have an intrinsic GTPase activity that is reduced by mutations corresponding to the transforming mutations in p21 (27,31). As in the case of their mammalian homologs, the *RAS1* and *RAS2* gene products are associated with the plasma membrane (11). The best illustration of similarity between the mammalian and yeast proteins is shown by the ability of the genes to function in their heterologous systems. Both the cellular *H-ras* gene (16) and its viral counterpart (8), *vHa-ras*, will substitute for the essential function of the yeast genes, and a modified yeast gene transforms NIH-3T3 cells (8).

The similarities between p21 and other GTP binding proteins (G proteins) (reviewed in (14)) and the subcellular location of p21 on the inner surface of the plasma membrane have led to the obvious hypothesis that p21 is involved in signal transduction, possibly mediating the signal

[1]Department of Biology/G5, University of Pennsylvania, Philadelphia, Pennsylvania 19104

37. Colloquium Mosbach 1986
Cell Cycle and Oncogenes

for a growth hormone. p21 is clearly not the G protein of several well-characterized signal transduction systems, and direct evidence for its role in modulating a signal for one or more of the growth factors is lacking. In yeast, however, recent evidence implicates *RAS* in modulating adenylate cyclase activity. As in mammalian cells, yeast adenylate cyclase is stimulated by GTP in the presence of Mg^{2+} (4). Toda et al. (32) have observed that yeast strains lacking *RAS2* have reduced adenylate cyclase activity, while strains containing a *RAS2* mutation analogous to the transforming mutations in p21 (*RAS2* ^{val19}) have high constitutive levels of adenylate cyclase activity. Thus, yeast *RAS* appears to behave as if it were a G protein required for activation of adenyate cyclase. Also in support of this model is the observation that a mutation which bypasses a defect in adenylate cyclase and results in loss of normal regulatory function of the cAMP-dependent protein kinase (*bcy1*) (26) also bypasses the normal requirement for *RAS*.

Phenotypes of both the dominant *RAS2*^{val19} mutation and recessive *RAS* disruptions implicate *RAS* with a role in the response to nutrient levels. Lillie and Pringle (18) have shown that starvation for ammonium ions, phosphate, sulfate, or glucose leads to rapid accumulation of the storage carbohydrates, glycogen and trehalose. Nitrogen starvation is required in most strains to initiate the meiotic cell cycle (sporulation). Strains carrying disruptions of *RAS2* sporulate on rich media (29,32) and hyperaccumulate glycogen and trehalose (29). In contrast, strains with *RAS2*^{val19} are unable to sporulate and accumulate glycogen (32).

Strains with disruptions of *RAS2* fail to grow effectively on nonfermentable carbon sources (10,29). The growth defect of *ras2* is somewhat different than most petite mutations. First, the defect is temperature-sensitive, allowing growth on ethanol or glycerol at 20°C, but not permitting growth on either carbon source at 37°C. We have reported that this defect of *ras2* can be eliminated by increased gene dosage of *RAS1*, either by supplying *RAS1* on an autonomously replicating plasmid (8) or as tandem copies integrated into the *RAS1* locus (29).

The growth defect of *ras2* reverts at a high frequency, giving rise to papillations at a frequency of one in 10^4-10^5 per cells plated on ethanol, glycerol, or acetate-containing media. In our initial report we characterized four independent revertants (29), finding that the suppression was due to mutations in one of three genes unlinked to either *RAS1* or *RAS2*. More recently, in an extensive analysis of 79 independent revertants (3), we find that mutations in two additional genes are responsible for the resumption of growth on nonfermentable carbon sources. In this report we summarize these recent results.

Results

Extragenic Suppressors of *RAS*

Most strains of *Saccharomyces cerevisiae* that lack a functional *RAS2* gene are unable to grow on media containing a nonfermentable carbon source (10,29). Although we do not fully understand the reason for this defect, it is at least in part due to differential expression of the *RAS* genes. Breviario et al. (2) have shown that the steady state level of *RAS1* messenger RNA is greatly reduced in cells growing on ethanol or acetate as the carbon source. A yeast strain lacking a functional *RAS2* gene would have insufficient *RAS* gene product when grown on media containing only a non-fermentable carbon source.

As we have previously noted (3,29), the gluconeogenic defect due to the lack of *RAS2* and in some cases the lethality due to the absence of both *RAS1* and *RAS2* can be suppressed by extragenic suppressors (called *sra* mutations). The mutations responsible for the reversion behave as Mendelian alleles, as shown by the 2:2 segregation for growth on ethanol when the revertants are mated to a *ras2* strain and subjected to tetrad analysis. In some cases the diploids heterozygous for a suppressor grow on ethanol or glycerol, indicating that the suppressor is dominant to its wild-type allele. With other suppressors the heterozygous diploids fail to grow on ethanol, indicating a recessive suppressor. Given the recent observation by Breviario et al. (2) that *RAS1* mRNA fails to accumulate on media containing ethanol or acetate as the sole carbon source, it stands to reason that the extragenic suppressors of this gluconeogenic defect could function by one of two mechanisms: either by allowing *RAS1* mRNA to accumulate on nonfermentable carbon sources, or by relieving the requirement for *RAS*. We have recently isolated and characterized 79 independent suppressor mutations, 68 of which have been assigned to one of five loci (3). Twenty two of these, in one recessive complementation group designated *sra6*, appear to work by allowing *RAS1* mRNA to accumulate on nonfermentable carbon sources (2). The other suppressors (*sra1*, *SRA3*, *SRA4* and *sra5*) appear to function by reducing or eliminating the requirement for *RAS1* and *RAS2* expression. These suppressors eliminate the normal essential requirement for *RAS* and also suppress the temperature-sensitive phenotypes of *cdc25* and *cdc35* mutations. Many of these suppressors have pleiotropic phenotypes, some of which are illustrated in Fig. 1.

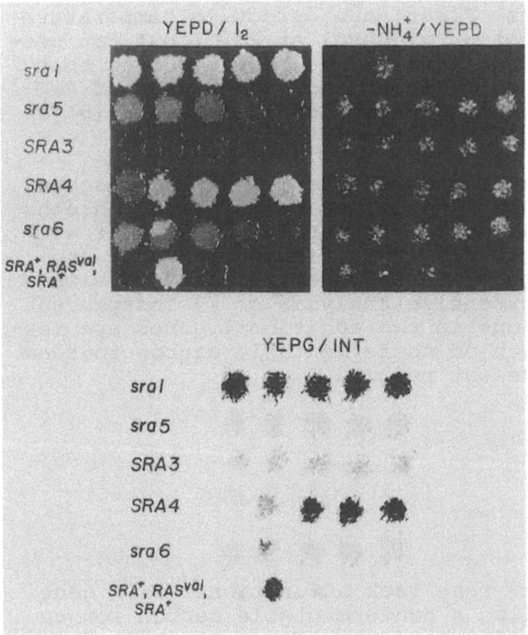

Fig. 1. Phenotypes associated with *sra* mutations. Strains containing different *sra* mutations were grown on rich media containing glucose as a carbon source (1% yeast extract 2% peptone 2% glucose, YEPD) and stained with iodine vapor (*top left*); incubated on ammonium sulfate-deficient media for 48 h and replica plated onto YEPD (*top right*); or grown on complete media containing glyerol as the sole carbon source and overlayed with the tetrazolium dye INT. Each row, except for the last, contains representative alleles of each *sra* loci. In the last row, a strain 301 (*MATα leu2 ura3 RAS2 ala 18val19 trp1 his3 lys1 lys2*) is flanked by colonies of JC302-26B (*MATa leu2 ras2-53o ura3 his4 SUC2*), the strain from which all the *sra* mutations were isolated

SRA1

To date 18 independent alleles, including two amber mutations, have been found in the recessive complementation group *sra1*. The *sra1* mutations are characterized by pleiotropic phenotypes. Strains with many *sra1* alleles fail to accumulate glycogen or trehalose, have a recessive sporulation defect, accumulate Zn-porphyrin, reduce tetrazolium dyes on nonfermentable carbon sources, and show unusually low viability on media deficient in ammonium sulfate. Some of these phenotypes are illustrated in Fig. 1, where congenic strains containing representative alleles for each of the *sra* loci are displayed. The pleiotropic phenotypes of many *sra1* mutants show allelic variation. In general, weaker alleles show the same trend in all traits. For example, alleles that only weakly reduce the tetrazolium dyes p-iodonitrotetrazolium violet (INT) and triphenyltetrazolium chloride (TTC), also accumulate glycogen (3). Another example of allelic variations is illustrated in Fig. 1. Strain 303-65 (*sra1-11*) (second *sra1* containing strain from the left) shows increased viability on ammonium sulfate deficient medium (Fig. 1, top right) relative to other *sra1* alleles but retains the ability to reduce INT (Fig. 1, bottom).

Many of the characteristics of *sra1* are also found for the dominant *RAS2* mutation, $RAS2^{ala18val19}$. A strain with this mutation (strain 301), like that with the previously reported $RAS2^{val19}$ mutation (32) fails to sporulate. Strain 301 also accumulates Zn porphyrin, and, as shown in Fig. 1, strongly reduces tetrazolium dyes and fails to accumulate glycogen. Strain 301 also shows reduced viability on ammonium sulfate-deficient media, but the loss of viability is not as rapid as with many *sra1* alleles.

Yeast cells normally require at least one of the two *RAS* genes for viability (15,28). Some alleles of *sra1* allow strains containing disruptions in both *RAS1* and *RAS2* to grow, thus bypassing the essential requirement for *RAS* (3,29). This is shown by crossing strains containing *ras2-530* (marked with the *LEU2* gene) and *sra1* alleles with strains containing a *RAS1* disruption (marked with *URA3*). With *sra1* present in the cross Leu^+Ura^+ clones are frequently observed. All such clones contain *sra1*, as scored by the failure to accumulate glycogen. Without the *sra1* allele no spore clones are recovered with the Leu^+, Ura^+ phenotype.

A duplication of one of the *RAS* genes, caused by a transposition event or by disomy for either chromosome XIV or XV (location of *RAS2* and *RAS1*, respectively) is one possible explanation for an apparent bypass phenotype for an *sra* mutation. If such a duplicated *RAS* gene segregates independently of the normal copy, it would be possible to recover haploids containing both *RAS* disruptions but kept alive by a third, nondisrupted *RAS* gene. We have ruled out this explanation by Southern hybridization analysis. Nitrocellulose filters, containing yeast DNA from tetrads in which disruptions of both *RAS* genes and *sra1* segregated, were hybridized with ^{32}P labeled *RAS1* and *RAS2* DNA. As shown in Fig. 2, spore clones that should have contained disruptions ob both *RAS1* and *RAS2* (RX23-3A in Fig. 2) contained no additional bands of hybridization. The same is true for *SRA3* (RX55-2C in Fig. 2) and for similar crosses with *SRA4* (data not shown).

In addition to allowing cells to grow without *RAS1* and *RAS2*, *sra1* alleles also suppress the temperature-sensitive phenotype of two previously characterized cell division cycle mutants. A detailed analysis of this suppression will be presented elsewhere (Robinson, Cannon and Tatchell, in preparation) but we summarize our conclusions below. *cdc25* and *cdc35* are two cell division cycle mutants that arrest as unbudded, nongrowing cells at their restrictive temperature. Like *ras2-530*, *cdc25* and *cdc 35* sporulate on rich media (25). *cdc35* and *cyr1* are probably both alleles of the adenylate cyclase structural gene (1,5,17,23). For these reasons we tested the ability of *sra-1* to suppress the temperature-sensitive phenotype of *cdc25* and *cdc35*.

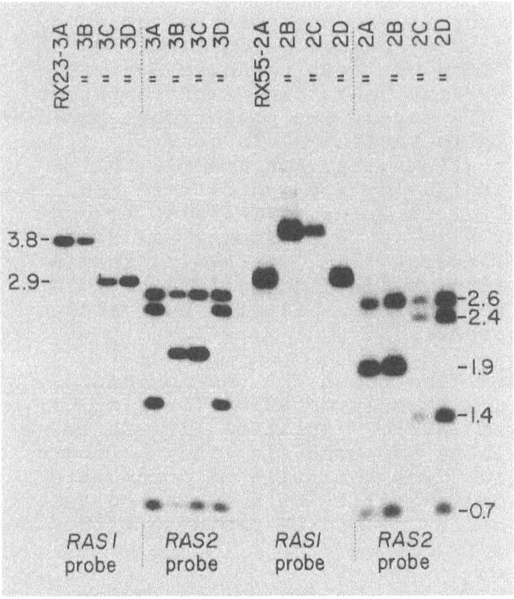

Fig. 2. Southern analysis of the *RAS* bypass mutants, *sra-1* and *SRA3-1*. DNA was prepared from tetrads derived from a cross between RX14-1C (*MATa leu2 ura3 lys2 trp1 ras 2-530 sra1-1)* and EG87-2B *(MATα leu 2 ura 3 lys2 ras1-545*) (RX23) and derived from a cross between RX8-2B *(MATa leu2 ura3 trp1 his4 lys2 ras2-530 SRA3-1)* and EG73-14D *(MATα leu2 ura3 lys1 his3 ras1-545)* (RX55). Diploid strains RX23 and RX55 are heterozygous for *ras1* and *ras2* disruptions and also heterozygous for either *sra1-1* and *SRA3-1*, respectively. In order to distinguish between wild-type and disrupted *RAS* genes, the DNA was digested with *Eco*RI and *Bam*HI and hybridized with a *RAS1* nick-translated probe or digested with *Cla*I and *Hind*III and hybridized with a *RAS2* probe. Numbers to the *right* and *left* of the autoradiograms refer to molecular weights of the bands in kilobases, determined using phageλ*Hind*III DNA fragments as standards. The 3.8 kb. and 2.9 kb. *RAS1*-specific bands correspond to the *RAS1* and *ras1-545* alleles respectively, while the 2.4 and 1.4 kb *RAS2* bands correspond to the *ras2-530* allele (28). Spore clones RX23-3A and RX55-2C show only the bands characteristic of the disruption alleles of both *ras1* and *ras2*. This genetic constitution is normally lethal but the presence of either *sra1-1* and *SRA3-1* allows the double mutants to grow

Tetrad analysis of diploids heterozygous for a *cdc* mutation normally shows two spore clones with the *cdc* phenotype (arrest at the restrictive temperature) and two wild-type spores. When the *sra1-1* mutation was present in the cross, however, growth at 37°C did not always segregate 2:2. In the case of *cdc25* and *cdc35*, there were always more spores without the temperature-sensitive phenotype than with it. In all cases, those spores with the *sra1-1* phenotype (failure to accumulate glycogen) always grew at 37°C. When spore clones that we reasoned should contain both the *cdc* mutation and *sra1-1* were backcrossed to a wild-type strain, haploid meiotic progeny were isolated that contained either the cdc gene or the *sra1-1* gene. Spore clones with the *sra1* phenotype were always able to grow at 37°C, irrespective of their *cdc* genotype. Our interpretation of these data is that *sra1* phenotype were always able to grow at 37°C, irrespective of their *cdc* genotype. Our interpretation of these data is that *sra1* suppresses both *cdc25* and *cdc35*. *sra1* did not suppress *cdc19*, *cdc33*, *cdc28* or *cdc36* (four other G_1 arrest mutants) nor any of a number of other *cdc* mutants tested.

We have found that protein kinase activity in extracts from *sra1-18* strains are insensitive to stimulation by cAMP, as assayed with histone or protamine substrates (3). We have recently found that the *sra1* gene codes for a protein with significant homology to bovine type 2 regulatory subunit of the cAMP-dependent protein kinase (Cannon and Tatchell in preparation). *bcy1*, a mutation isolated as a suppressor of adenylate cyclase mutations (19) and recently shown to bypass the requirement for *RAS* (32), is probably allelic to *sra1*.

SRA4

Seven dominant suppressors are tightly linked to *met3* and have been designated as alleles of *SRA4*. Many *SRA4* alleles share the pleiotropic phenotype of *sra1*. They accumulate little or no glycogen, accumulate high levels of Zn-porphyrin, and strongly reduce tetrazolium dyes (see Fig.1). They also bypass the normal requirement for *RAS*, allowing *ras1 ras2* double mutants to grow (data not shown). Like *sra1*, *SRA4* alleles show allelic variation. As presented in Fig. 1, strains containing *SRA4-7* (left most *SRA4* colony in Fig. 1) accumulate glycogen and do not bypass the requirement for both *RAS* genes. Strains containing stronger alleles of *SRA4* (the four *SRA4* colonies to the right in Fig. 1) fail to accumulate glycogen and do confer *RAS* independence.

SRA4 is tightly linked to the structural gene for adenylate cyclase (*cdc35/cyr1*). We find that strains containing *SRA4* alleles have a higher specific activity of adenylate cyclase than congenic wild-type strains (3). These data support the hypothesis that *SRA4* may be a dominant allele of *cdc35/cyr1*. Kataoka et al. (17) report that some insertion mutations in the cloned *CYR1* gene behave like our *SRA4* mutations, giving high *RAS* -independent levels of adenylate cyclase activity.

sra5

Seven suppressor mutations fall into a complementation group designated *sra5*. Biochemical analysis of two strains carrying alleles of *sra5* reveal significantly reduced levels of cAMP phosphodiesterase. JC303-16 (*sra5-3 RAS1 ras2-530*) possessed 80-fold reduced levels of a high affinity phosphodiesterase relative to the congeneic parent strain JC302-26B. This reduced activity segregated with *sra5* in the three tetrads analyzed (3). *sra5* suppressors are not strictly bypass mutants like *sra1* and *SRA4*. None of the *sra5* mutations, including two nonsense alleles, allow spores lacking a functional *RAS* gene to grow on rich media (2% peptone, 1% yeast extract, 2% glucose). Often, *ras1, ras2 sra5* spores germinate on rich media and divide one to three times before arresting. Growth can be maintained by including cAMP in the media (R.B. Wilson unpublished observations). Surprisingly, we find that *sra5 ras1 ras2* strains will grow in the absence of exogenous cAMP on minimal media (R.B. Wilson unpublished observation). A similar behavior is observed in the ability of *sra5* to suppress the temperature sensitivity of *cdc25* and *cdc35*. *sra1*, *SRA3* and *SRA4* allow strains containing *cdc25* or *cdc35* (*SRA4* cannot be tested with *cdc35* since the two are probably allelic) to grow at temperatures well above the restrictive temperature of the *cdc* mutants. *sra5*, on the other hand, raises the restrictive temperature to only 35°C; the mutants fail to grow above 37°C.

SRA3

Many *sra1* and *SRA4* mutants fail to accumulate glycogen, are unable to sporulate, and have increased ability to reduce tetrazolium dyes. Sixteen dominant suppressors, all mapping 4 map units from *ino1*, allow *ras1 ras2* cells to grow yet do not have the other traits that accompany the other bypass suppressors. As shown in Fig. 1, *SRA3* containing strains stain very dark with iodine, indicating high levels of glycogen accumulation in these strains. All *ras1 ras2 SRA3* strains have severely reduced growth rates, which indicates that *SRA3* does not completely bypass the *RAS* requirement.

Discussion

We have isolated mutations in at least five genes that suppress the inability of *ras2* mutants to grow on nonfermentable carbon sources. One of these, *sra6*, alters the expression of *RAS1* gene and allows *RAS1* mRNA to accumulate on nonfermentable carbon sources. The other four supressors function by reducing, and in many cases eliminating, the essential requirement for *RAS1* or *RAS2* genes and allow strains without either *RAS1* or *RAS2* to grow and divide. We have accumulated evidence implicating at least three of these genes in the metabolism of cAMP, as outlined in Fig. 3. The *SRA4* mutations map to the adenylate cyclase locus and show increased adenylate cyclase activities. *sra5* mutants have decreased levels of cAMP phosphodiesterase, suggesting that *sra5* is either the structural gene for a phosphodiesterase or a regulator thereof. *sra1* codes for the regulatory subunit of cAMP-dependent protein kinase.

The genetic and biochemical analysis of our *sra* mutations corroborates previous studies showing that the *RAS* genes are required for normal adenylate cyclase activity. The simplest interpretation of both the genetic and biochemical data is that the *RAS* gene products directly stimulate adenylate cyclase in a manner analogous to the stimulation of mammalian adenylate cyclase by the G or N proteins (14). cAMP, in turn, has a key role in regulating cell growth in yeast ((21) reviews these data in detail). Cells that lack the ability to synthesize cAMP, i.e., that contain mutations in *RAS* or adenylate cyclase *(cdc35/cyr1)*, arrest in G1 of the cell cycle as unbudded, nongrowing cells. The main function of cAMP, in turn, appears to be the activation of the cAMP-dependent protein kinase, which it does by binding to the inhibitory subunit *(bcy1/sra1)*, thereby releasing the active catalytic subunit. Presumably phosphorylation of specific substrates by this kinase is a requirement for cell growth. We have shown that the essential function for *RAS* in the mitotic cell cycle can be bypassed by mutations that either increase adenylate cyclase activity *(SRA4)*, inhibit the degradation of cAMP *(sra5)* or increase cAMP-dependent protein kinase activity *(sra1)*. One might have expected *sra* mutations in the kinase catalytic

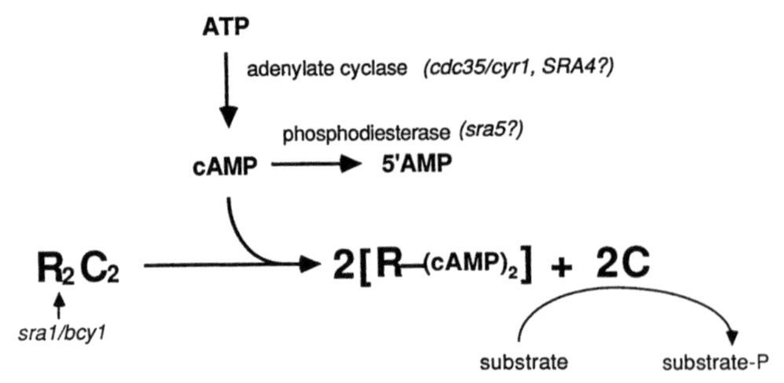

Fig. 3. Possible involvement of SRA genes in cAMP metabolism

subunit gene or in a particular substrate for the kinase (assuming there is only one essential substrate). Such functions are possible for *SRA3* or for some of our remaining uncharacterized dominant *SRA* mutations. The cloning and molecular characterization of many of our suppressors, now underway in the laboratory, should confirm or rule out our hypothesis concerning their function.

The cAMP regulatory system appears to be tuned to the nutritional state of the cell. Although it has not been rigorously shown that starvation directly inhibits adenylate cyclase, it is known that loss of nitrate, sulfate, phosphate, or a carbon source leads to rapid accumulation of storage carbohydrates (18), which are known to be partially regulated by the cAMP-dependent protein kinase (22,33,34). Regulation of entry into the meiotic cell cycle also appears to be regulated by cAMP metabolism. In wild-type yeast strains sporulation is usually induced by starvation conditions. However, strains with defects in adenylate cyclase or *RAS* activity *(ras2, cdc35/cyr1)* sporulate on rich media. Conversely, strains that contain *RAS2*^{val19}, *bcy1* or *SRA4* fail to sporulate.

RAS and adenylate cyclase do not affect the cell cycle per se but primarily affect cell growth. Specific control of the cell cycle is regulated by another set of genes, identified by the class I *(cdc28)* class of cell cycle mutants, which cause cell cycle arrest at the restrictive temperature without affecting cell growth (24). This class of cell division cycle mutants is not suppressed by *sra* mutations. Clearly, the systems of growth control and cell cycle control must be integrated at some point. A signal or signals, indicating that the nutritional requirements necessary for growth have been met, is likely to be one of the preconditions required for cells to enter a new cell cycle. Perhaps a phosphorylation event, mediated by the cAMP-dependent protein kinase, is one such signal.

References

1. Boutelet F, Petitjean A, Hilger F (1985) Yeast *cdc35* mutants are defective in adenylate cyclase and are allelic with *cyr1* mutants whilse *CAS1*, a new gene, is involved in the regulation of adenylate cyclase. EMBO J 4:2635-2641
2. Breviario D, Hinnebusch A, Cannon J, Tatchell K, Dhar R (1986) Carbon source regulation of RAS1 expression in *Saccharomyces cerevisiae* and the phenotypes of ras2-cells. Proc Natl Acad Sci USA 83:4152-4156
3. Cannon JF, Gibbs JB, Tatchell K (1986) Suppressors of the ras2 mutation of *Saccharomyces cerevisiae*. Genetics 113:247-264
4. Casperson GF, Walker N, Brasier AR, Bourne HR (1983) A guanine nucleotide-sensitive adenylate cyclase in the yeast *Saccharomyces cerevisiae*. J Biol Chem 258:7911-7914
5. Casperson GF, Walker N, Bourne HR (1985) Isolation of the gene encoding adenylate cyclase in *Saccharomyces cerevisiae*. Proc Natl Acad Sci USA 82:5060-5063
6. Dahr R, Nieto A, Koller R, DeFeo-Jones D, Scolnick EM (1984) Nucleotide sequence of two *ras*^{H}*-related* genes isolated from the yeast *Saccharomyces cerevisiae*. Nucleic Acids Res 12:3611-3618
7. DeFeo-Jones D, Scolnick EM, Koller R, Dhar R (1983) *ras*-related gene sequences identified and isolated from *Saccharomyces cerevisiae*. Nature (London) 306:707-709
8. DeFeo-Jones D, Tatchell K, Robinson LC, Sigal I, Vass W, Lowry DR, Scolnick EM (1985) Mammalian and yeast *ras* gene products: biological function in their heterologous systems. Science 228:179-184
9. Ellis TE, Lowy DR, Scolnick EM (1982) The viral and cellular p21 *(ras)* gene family. Adv Viral Oncol 1:107-126
10. Fraenkel D (1985) On *ras* gene function in yeast. Proc Natl Acad Sci USA 82:4740-4744

11. Fujiyama A, Tamanoi F (1986) Processing and fatty acid acylation of *RAS1* and *RAS2* proteins in *Saccharomyces cerevisiae*. Proc Natl Acad Sci USA 83:1266-1270
12. Gallwitz D, Donath C, Sander C (1983) A yeast gene encoding a protein homologous to the human C-has/bas proto oncogene product. Nature (London) 306:704-707
13. Gibbs JB, Sigal IS, Scolnick EM (1985) Biochemical properties of normal and oncogenic ras-p21. Trends Biochem Sci 10:350-353
14. Gilman AG (1984) G proteins and dual control of adenylate cyclase. Cell 36:577-579
15. Kataoka T, Powers S, McGill C, Fasano O, Strathern J, Broach J, Wigler M (1984) Genetic analysis of yeast *RAS1* and *RAS2* genes. Cell 37:437-445
16. Kataoka T, Powers S, Cameron S, Fasano O, Goldfarb M, Broach J, Wigler M (1985) Functional homology of mammalian and yeast *RAS* genes. Cell 40:19-26
17. Kataoka T, Broek D, and Wigler M (1985) DNA sequence and characterization of the *S. cerevisiae* gene encoding adenylate cyclase. Cell 43:493-505
18. Lillie SH, Pringle JR (1980) Reverse carbohydrate metabolism in *Saccharomyces:* responses to nutrient limitation. J Bacteriol 143:1384-1394
19. Matsumoto K, Uno I, Oshima Y, Ishikawa T (1982) Isolation and characterization of yeast mutants deficient in adenylate cyclase and cyclic AMP dependent protein kinase. Proc Natl Acad Sci USA 79:2355-2359
20. Matsumoto K, Uno I, Ishikawa T (1983) Initiation of meiosis in yeast mutants defective in adenylate cyclase and cyclic AMP dependent protein kinase. Cell 32: 417-423
21. Matsumoto K, Uno I, Ishikawa T (1985) Genetic analysis of the role of cAMP in yeast 1:15-24
22. Ortiz CH, Maia JCC, Tenan MM, Braz-Padrao GR, Mattoon JR, Panek AD (1983) Regulation of yeast trehalase by a monocyclic AMP-dependent phosphorylation-dephosphorylation cascade system. J Bacteriol 153:644-651
23. Powers S, Kataoka T, Fasano O, Goldfarb M, Broach J, Wigler M (1984) Genes in *S. cerevisiae* encoding proteins with domains homologous to the mammalian *ras* proteins. Cell 36:607-612
24. Pringle JR, Hartwell LH (1982) The *Saccharomyces cerevisiae* cell cycle. In: Strathern J, JOnes E, Broach J (eds) The molecular biology of the yeast *Saccharomyces*-life cycle and inheritance. Cold Spring Harbor Lab, Cold Spring Harbor, New York, pp 97-142
25. Shilo V, Shimchen G, Shilo B (1978) Initiation of meiosis in cell-cycle initiation mutants of *Saccharomyces cerevisiae*. Exp Cell Res 112:241-248
26. Tamanoi F, Walsh M, Kataoka T, Wigler M (1984) A product of yeast *RAS2* is a guanine nucleotide binding protein. Proc Natl Acad Sci USA 81:6924-6928
27. Tamanoi F, Samiy N, Rao M, Walsh M (1985) Enzymatic properties of yeast *RAS2* protein. In: Feramisco J, Ozanne B, Stiles L (eds) Cancer cells, vol. III. Growth factors and transformation. Cold Spring Harbor Lab, Cold Spring Harbor New York, pp 251-256
28. Tatchell K, Chaleff DT, DeFeo-Jones D, Scolnick EM (1984) Requirement of either of a pair of *ras*-related genes of *Saccharomyces cerevisiae* for spore viability. Nature (London) 309:523-527
29. Tatchell K, Robinson LC, Breitenbach M (1985) *RAS2* of *Saccharomyces cerevisiae* is required for gluconeogenic growth and proper response to nutrient limitation. Proc Natl Acad Sci USA 82:3785-3789
30. Temeles GL, DeFeo-Jones D, Tatchell K, Ellinger MS, Scolnick EM (1984) Expression anc characterization of *ras* mRNA's from *Saccharomyces cerevisiae*. Mol Cell Biol 4:2298-2305
31. Temeles GL, Gibbs JB, D'Alonzo JS, Sigal IS, Scolnick EM (1985) Yeast and mammalian *ras* proteins have conserved biochemical properties. Nature (London) 313: 700-703
32. Toda T, Uno I, Ishikawa T, Powers S, Kataoka T, Broek D, Cameron S, Broach J, Matsumoto K, Wigler M (1985) In yeast, *RAS* proteins are controlling elements of adenylate cyclase. Cell 40:27-36
33. Uno I, Matsumoto K, Adachi K, Ishikawa T (1983) Genetic and biochemical evidence that trehalase is a substrate of cAMP-dependent protein kinase in yeast. J Biol Chem 258:10867-10872
34. Wingender-Drissen R, Becker JU (1983) Regulation of yeast phosphorylase by phosphorylase kinase and cAMP-dependent protein kinase. FEBS Lett 163:33-36

Functional Analysis of the *ras*-Related *YPT1* Gene Product in Yeast

D. Gallwitz, R. Brökel, C. M. T. Molenaar, A. J. G. Rauh, H. D. Schmitt, and P. Wagner[1]

Introduction

H-ras, *K-ras* and *N-ras* are members of a family of highly homologous vertebrate proto-oncogenes that code for a protein of 189 amino acids. The p21 protein products of the viral *v-H-ras* and *v-K-ras* genes are the causing agents of rat sarcomas and mutated forms of all three *ras* proto-oncogenes have been shown to be contained in several human tumours and tumour cell lines and to be able to transform malignantly NIH3T3 fibroblasts ((1,2) for review).

The p21 *ras* gene products are bound to the inner surface of the plasma membrane, they bind GTP and GDP specifically and possess intrinsic GTPase activity ((3) for review). Although the *ras* proteins have been well characterised biochemically, their normal cellular function has not been disclosed. Therefore, the discovery in yeast of genes with significant homology to the vertebrate *ras* genes (4,5,6) raised the hope that, with the help of sophisticated genetic analyses possible with this unicellular organism the cellular function of *ras* proteins might come to light. To some extent these expectations have already been fulfilled as it is now clear that one set of two functionally related yeast *ras* genes, *RAS1* and *RAS2*, exerts its (major) function by regulating adenylate cyclase activity. Recently we have identified in the yeast *S. cerevisiae* a *ras*-related gene, *YPT1*, which resides on chromosome 6 between the actin and the β-tubulin genes (4). Over a region of 165 amino acids the 206-residue-long *YPT* protein displays a 38% homology with the 189 amino acids of the human *c-ras* gene products implying very similar secondary and tertiary structure. In contrast to the yeast *RAS1/RAS2* gene products that display even higher homology to the mammalian ras proteins, the *YPT* protein seems to be required for mitosis and the proper microtubule organisation and function (11).

Functional Analyses of the YPT1 Gene Product

The ease with which a wild-type allele can be replaced with a mutant allele by homologous recombination using linear DNA fragments makes the unicellular yeast an ideal organism to study the function of normal gene products and the consequences of mutation in them. To analyse whether the *YPT1* gene product is essential for cell viability, a gene disruption experiment was performed as schematically outlined in Fig. 1. The protein-coding region of the gene was interrupted with a yeast *LEU2* marker gene flanked by bacterial plasmid sequences. In diploid *leu2⁻* yeast cells one *YPT1* allele was replaced with the disrupted gene,

[1]Physiologisch-Chemisches Institut, Universität Marburg, Emil-Mannkopff-Str. 1, 3550 Marburg, FRG

37. Colloquium Mosbach 1986
Cell Cycle and Oncogenes

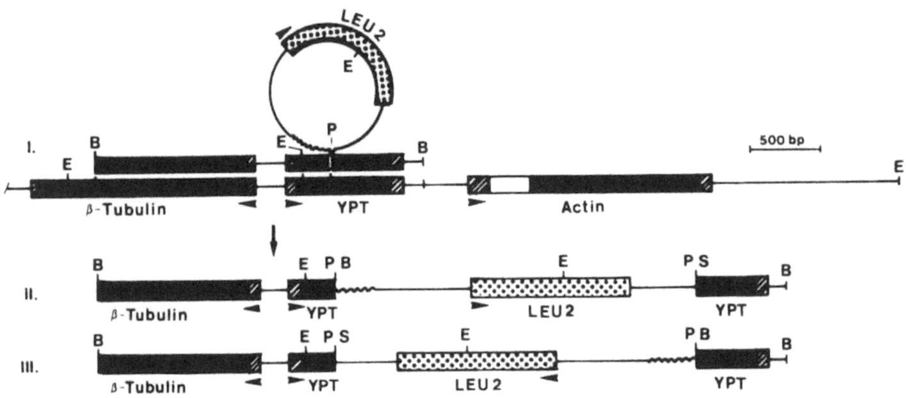

Fig. 1. Schematic representation of *YPT1* gene disruption. A diploid yeast strain was transformed with a BamHI fragment containing the *YPT1* genes disrupted by the yeast LEU2 gene. The two possible orientations of the gene disrupting fragment, shown in II and III, were tested in the disruption experiment and gave the same result. *YPT1* gene disruption on one of the two chromosomes VI was verified by Southern blot analysis using *Bam*HI and *Eco*RI digested cellular DNA and a ^{32}P-nick translated *YPT1* specific DNA fragment as hybridisation probe. ～ pBR322 sequence. B, *Bam*HI; Bg, *Bg*III; E, *Eco*RI; H, *Hin*dIII; P, *Pst*I, S, *Sal*I. *Small arrows* indicate direction of transcripiton

and the transformed cells were subjected to spore analysis. It was found that the spores with the *YPT1* null allele could germinate but were only able to go through one or two cell divisions. This experiment clearly established that the *YPT1* gene product is required for cell viability and that the *RAS1* and *RAS2* proteins are not able to complement the *YPT1* defect (11).

Our initial attempt to search for a phenotype following *YPT1* loss exploited the finding of Murray and Szostak (12) that circular plasmids containing chromosomal replication origins (ARS) have a strong tendency to segregate with the mother cell at mitosis. Loss of *YPT1* function was followed in a haploid yeast strain with a disrupted *YPT1* gene on chromosome VI and a functional *YPT1* gene on the autonomously replicating plasmid YRp7. By micromanipulation the progeny of individual cells was followed after separating mother and daughter cells for several generations. Plasmid-bearing cells could then be identified by their ability to form visible colonies. After plasmid (and *YPT1* gene) loss, cells could go through several divisions and usually microcolonies of three to 70 cells were observed (Fig. 2). This was obviously due to the presence of enough *YPT* protein derived from the intracellularly amplified wild-type gene on the plasmids allowing three to six rounds of cell division. Inspection of the *YPT*- deficient and arrested cells showed no uniform phenotype, but a significant number of cells had two buds and were abnormally large.

To study the arrest phenotype in large populations of cells we brought the *YPT1* gene under the control of the galactose-inducible *GAL10* promoter and replaced the wild-type *YPT1* gene with the mutant allele on chromosome VI. By replacing the growth medium containing galactose (*YPT1* gene switched on) with medium containing glucose, the expression of the *YPT1* gene was turned off and cells stopped growing after about 11 h. Here again, cells arrested for more than 15 h increased in size and roughly 50% of them had two buds (11).

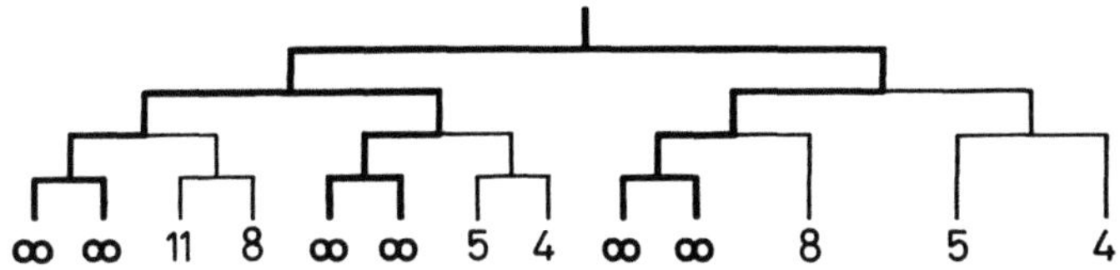

Fig. 2. Example of plasmid segregation in a haploid yeast clone carrying a functional *YPT1* gene as part of an autonomously replicating plasmid. The *YPT1* copy on chromosome VI was destroyed in the same way as outlined in Fig. 1. Plasmid loss, therefore, led to inviability of cells. Numbers indicate the amount of cells per microcolony after another 36 h of growth at 30°C (∞, visible colony). *Thick lines* show the inheritance of plasmids

The neighbourhood of the *YPT1* and the β-tubulin genes on the chromosome, the regulation of the two divergently transcribed genes from overlapping DNA sequences and the ability of both gene products to bind GTP prompted us to study the distribution of tubulin in the arrested cells. In yeast cells microtubules originate from the spindle pole body (SPB) embedded in the nuclear envelope and extend into the cytoplasm and into the nucleus, where some attach to the chromosomes and others meet the SPB on the opposite side of the nucleus to form the spindle. After switching off the *YPT1* gene transcription the microtubules, observed by indirect immunofluorescence using yeast anti-tubulin, appeared normal as long as the cell growth was unaffected (8 h). At the time the growth curve started to level off (11 h), mitotic spindles were unevenly stained and the cytoplasmic microtubules became abnormally long as if the assembly/disassembly of microtubules was impaired. At later times (20 h and more) microtubules were completely distorted and nuclear structures, identified by simultaneously staining the cellular DNA with DAPi, seemed to be completely disrupted (Fig. 3). Surprisingly, cells arrested for 20 h and more were, to about 80%, able to resume growth after turning on the *YPT1* gene in galactose-containing medium (11).

It should be pointed out that the effects on microtubule organisation and nuclear structure in *YPT*-depleted cells are not simply due to the growth arrest, since stationary cells, doubly labelled with DAPI and antitubulin, have (1) a clearly visible nucleus and (2) a small, dot-like microtubule-organising centre connected with the nucleus.

An obvious interpretation of the findings described would be that the *YPT* protein in yeast is directly involved in an organised growth and functioning of microtubules. However, the complete distortion of microtubule structures could also be an indirect consequence of the loss of *YPT* function.

In favour of the first interpretation is a dominant-lethal mutation ($Asn^{121} \rightarrow Ile$ substitution) that we created in the course of a mutational analysis of the *YPT* protein. The mutant gene under the control of the *GAL10* promoter could be turned on and off at will and it was observed that the production of the mutant protein led to growth arrest, loss of cell viability and to an non-uniform arrest phenotype with about one third of the cells having a small bud and two clearly separated nuclei residing in the mother cell and connected by an abnormal spindle. The spindle was thin, often pressed against the cell wall and it lacked cytoplasmic microtubules (Fig. 4).

Although the arrest phenotype of this mutation differs qualitatively from that observed in cells depleted in *YPT* protein, the perturbation of the mitotic spindle again hints at the importance of the *YPT* protein for a proper functioning of microtubule structures.

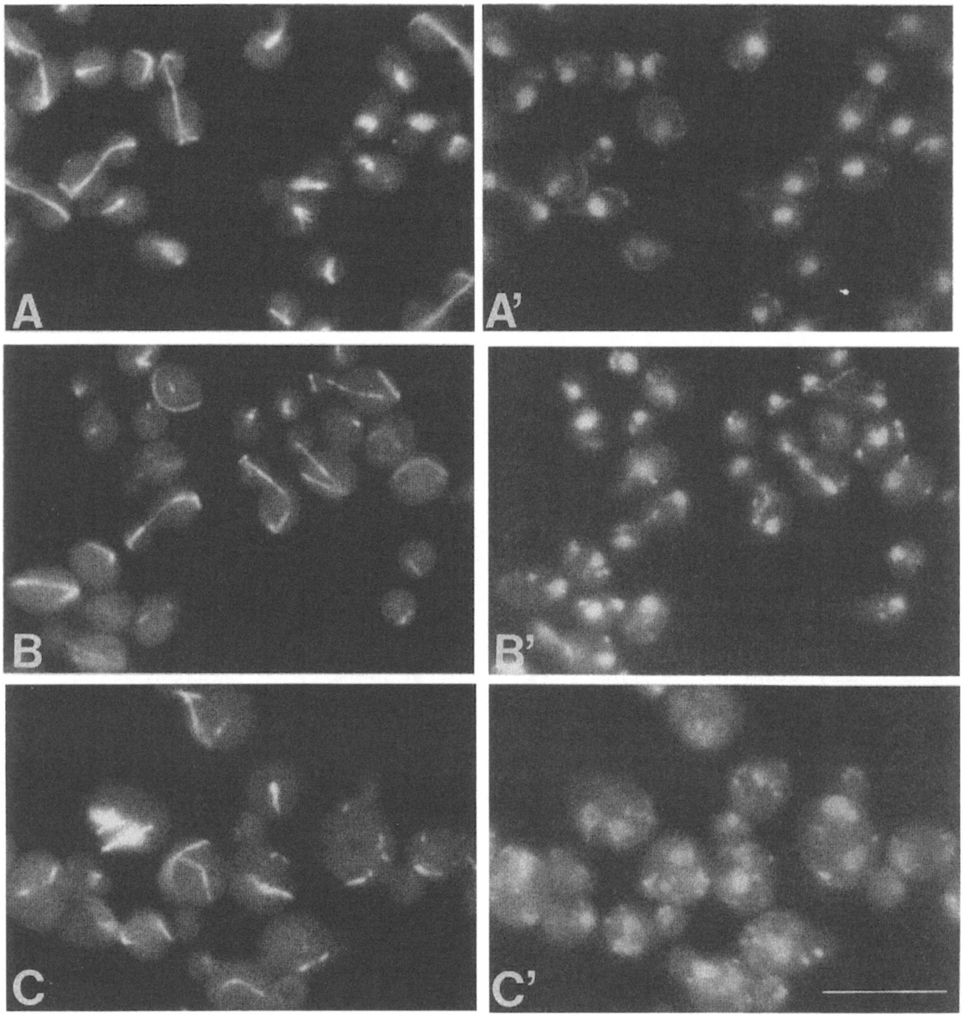

Fig. 3. Immunofluorescent staining of tubulin (A-C) and DNA (A'-C') of logarithmically growing cells (A) and cells in which *YPT1* gene transcription was turned off for 11 h (B) and 24 h (C). Note normal spindles extending between the separated nuclei in (A), and the beginning and complete distortion of microtubules in *YPT* protein deficient cells (B,C). Also notice the enlargement of arrested cells and the apparent absence of clear nuclear contours (C'). Antiserum against yeast tubulin was kindly provided by J.V. Kilmartin

The $Asn^{121} \rightarrow Ile$ substitution, which results in an inability of the mutant protein to bind GTP, might lead to a "locked in" conformation and an an irreversible binding of the protein to a physiological effector protein. This mutation offers the possibility to apply both genetic and biochemical methods to the search for cellular components interacting with the *YPT* protein.

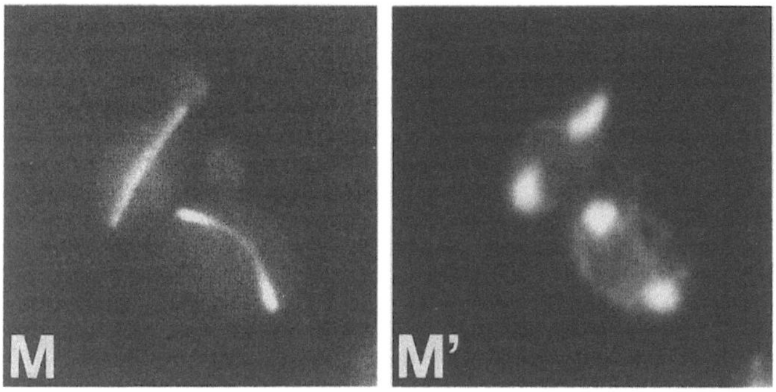

Fig. 4. Cells arrested following expression of the dominant lethal YPT^{ile121} allele. Staining with anti-tubulin (*M*) and DAPI (*M'*) shows two clearly separated nuclei in the mother cells that are connected by a thin spindle

```
                                     G    20 M
YPT      M N S E Y D Y L F K L L L I G N S G V G K S C L L
C-RAS^H          M T E Y K L V V V G A G G V G K S A L T

                                   40
YPT      L R F S D D T Y T N D Y I S T I G V D F K I K T V
C-RAS^H  I Q L I Q N H F V D E Y D P T I E   D S Y R K Q V

                        60         T
YPT      E L D G K T V K L Q I W D T A G Q E R F R T I T S
C-RAS^H  V I D G E T C L L D I L D T A G Q E E Y S A M R D

               80                                     100
YPT      S Y Y R G S H G I I I V Y D V T D Q E S F N G V K
C-RAS^H  Q Y M R T G E G F L C V F A I N N T K S F E D I H

                                                120
YPT      M W L Q E I D R Y A T S T   V L K L L V G N K C D
C-RAS^H  Q Y R E Q I K R V K D S D D V P M V L V G N K C D

                                  140
YPT      L K D K R V V E Y D V A K E F A D A N K M P F L E
C-RAS^H  L A A   R T V E S R Q A Q D L A R S Y G I P Y I E

                             160
YPT      T S A L D S T N V E D A F L T M A R Q I K E S M S
C-RAS^H  T S A K T R Q G V E D A F Y T L V R E I R Q H   K

                 180
YPT      Q Q N L N E T T Q K K E D K G N V N L K G Q S L T
C-RAS^H  L R K L N P P D E S G P G C M S C K C V L S

        200
YPT      N T G G G C C
```

Fig. 5. Comparison of the amino acid sequences of the human *c-H-ras* and the yeast *YPT* proteins. Identical residues in both proteins are boxed. Residues changed by site-directed mutagenesis are shown in **bold-faced letters**

Biochemical Properties of the YPT Protein

A comparison of the amino acid sequence of the yeast *YPT* and the human *c-H-ras* proteins is given in Fig. 5. Regions with perfect homology over five and more consecutive residues are sequence motifs that are also conserved in other GTP/GDP binding proteins (13,14,15).

To show that the *YPT* protein is a nucleotide-binding protein, total yeast cellular protein was separated by sodium dodecyl sulphate-polyacrylamide gel electrophoreses (SDS-PAGE) and, after removing the SDS from the gels, electrophoretically transferred to nitrocellulose filters. The *YPT* protein, identified as a 23.5 kDa protein using antibodies against an MS2 polymerase/ *YPT* fusion protein (11), could be shown to bind ^{32}P-GTP or ^{32}P-dGTP specifically.

	Mutation	Growth	GTP Binding	GTPase	Auto-phosph.
1)	Ser^{17} → Gly	+	+↓	+↑	-
2)	Lys^{21} → Met	lethal	-		
3)	Ala^{65} → Thr	+	+↓	+↓	+
4)	Asn^{121} → Ile	dominant lethal	-		
5)	Ser^{17} → Gly Ala^{65} → Thr	lethal§	+↓	+↓	+↓
6)	Δ Cys^{205,206}	lethal	+		

Fig. 6. Physiological and biochemical effects of mutations introduced into the *YPT* protein. By homologous recombination one *YPT1* wild-type allele of a diploid strain was replaced with the mutant allele and the growth properteis of the different haploid mutants were investigated after sporulation. GTP binding, GTPase and autophosphorylating activities were tested with bacterially produced proteins. *Arrows* indicate increase (↑) or decrease (↓) of activity. Mutant (5) formed microcolonies with up to 50 cells. In mutant (6) the two C-terminal cysteine residues are deleted.

Mutations were introduced in the *YPT1* gene to exchange single amino acid residues that were believed to be of functional importance with respect to the GTP binding capacity and GTPase activity of the protein (see Fig. 5). The mutated genes were used to replace the *YPT1* wild-type gene on chromosome VI to study the physiological effects of the mutant proteins in yeast cells. In parallel, the wild-type and mutant *YPT* proteins were also expressed in *E. coli* and purified by ion-exchange chromatography and gel filtration. The bacterially produced *YPT* protein was soluble and retained its GTP-binding capacity. It also had GTPase activity.

A summary of different *YPT* mutant proteins studied and their effects on cellular growth and biochemical properties is presented in Fig. 6. Substitutions Lys^{21}→Met and Asn^{121}→Ile were recessive and dominant lethal, respectively. According to the filter binding assay, these mutant proteins had lost their GTP-binding activity. In the light of recent X-ray crystallographic studies of the E. coli elongation factor EF-Tu (16) the conserved lysine-42 residue, corresponding to that in position 17 of *YPT* and to lysine-12 of the mammalian *ras* proteins, seems to form ionic contacts to the β-phosphate group of the bound GDP. Asn^{135}, corresponding to the asparagine residues in positions 121 and 116 of the *YPT* protein and mammalian *ras* proteins, respectively, seems to interact with the ketooxygen at C6 of the guanine ring (16, 22). It was therefore satisfying to find the change of these two conserved residues in the *YPT* protein to severely perturb the binding of GTP.

Human *ras* protiens can acquire transforming activity through different mutations of either glycine-12 (17,18) or glutamine-61 (19,20). These mutations decrease the intrinsic GTPase activity of the *ras* proteins and it is still an open question whether the impairment of GTPase activity is causally related to the transforming activity of the mutant *ras* proteins.

The *YPT* protein has a serine residue in position 17 which corresponds to Gly^{12} of the mammalian *ras* proteins. The substitution $Ser^{17} \rightarrow Gly$ in *YPT* resulted in an apparent decrease of the GTP-binding capacity and a two- to threefold increase in GTPase activity. The cellular growth was not affected by this mutation.

Ala^{65} of the *YPT* protein, corresponding to Ala^{59} in vertebrate *ras* proteins, was changed to threonine as this substitution, observed in viral *ras* proteins, was found to be responsible for the autophosphorylating activity of these mutated proteins. Thr^{59} is the acceptor site in the autophosphorylation reaction (21) because it seems to be close to the γ-phosphate group of the bound GTP (16,22). In analogy to the *ras* proteins (23), the $Ala^{65} \rightarrow Thr$ substitution conferred on the *YPT* protein autophosphorylating activity and, in addition, the apparent capacity to bind GTP and the GTPase activity of the mutant *YPT* protein decreased. Despite the changed biochemical properties of YPT^{thr65}, haploid cells carrying the mutant gene instead of the wild-type *YPT* gene were not influenced with respect to their growth properties. A double mutant with the substitutions $Ser^{17} \rightarrow Gly$ and $Ala^{65} \rightarrow Thr$ was lethal and cells with the mutant gene were able to go through only a limited number of divisions.

Taken together, these and other mutations that we have introduced into the *YPT* protein support the notion that the *YPT* protein and the mammalian *ras* proteins have very similar, if not identical biochemical properties.

Conclusions

In the hope of aiding an understanding of normal *ras* function in mammalian cells, 2 years ago several groups set out to define the cellular function of *ras*-like genes in the yeast *Saccharomyces cerevisiae* (4,5,6). Two sets of biochemically very similar but functionally different proteins, *RAS1/RAS2*, on the one hand, and *YPT1*, on the other, could be distinguished in yeast and a third group of genes with significant homology to the mammalian *ras* genes, RHO1/RHO2, has recently been discovered in this unicellular organism (P. Madaule, personal communication).

From a protein chemical point of view, all of the *ras*-like yeast proteins are expected to share very similar secondary and tertiary structure with the mammalian *ras* proteins. This obviously is a reflection of the prominent biochemical properteis of these proteins, their ability to bind and hydrolyze GTP. It seems likely that the *ras* proteins are part of signal transducing systems in which they communicate, analogous to the G-proteins in higher cells, with differenct receptors and/or effectors. Critical to the function of all *ras*-related proteins is the reversible binding of GTP, but it is clear that the different sets of *ras* proteins differ from each other by their functional specificity.

The major function of the *RAS1/RAS2* proteins in *S. cerevisiae* is to regulate adenylate cyclase activity (7,8). The *YPT* protein seems to be involved, directly or indirectly, in promoting proper microtube organisation and function (11). The *RHO* proteins might be integrated still in other regulation pathway(s).

Modified versions of human *ras* genes can partially replace *RAS1/RAS2* function in yeast (24,25) but other than in yeast, *ras* proteins in mammalian cells do not regulate adenylate cyclase activity. It is also worth mentioning that the inactivtation of the single *ras1* gene of the fission yeast *Schizosaccharomyces pombe* is not lethal and perturbs mating but not cAMP metabolism (26).

From the work with yeast *ras*-related genes we have learned that (1) proteins very similar in structure and almost identical in biochemical properties are functionally different and (2) what is true for yeast might not necessarily be true for mammalian cells. Nevertheless, we can anticipate learning a lot more about gene products in yeast that communicate with *ras* proteins and, heading in this direction, might, in the end, turn out to be a profitable way to an understanding of the involvement of *ras* proteins in normal and disordered cell proliferation.

Acknowledgements. We thank Petra Teutsch and Angela Fiebiger for expert technical assistance. The authors' work was supported by grants to D.G. from the Deutsche Forschungsgemeinschaft and the Fonds der Chemischen Industrie.

References

1. Bishop JM (1985) Trends Genet 1:245-249
2. Levinson AD (1986) Trends Genet 2:81-85
3. Gibbs JB, Sigal IS, Scolnick EM (1985) Trends Biochem Sci 10:350-353
4. Gallwitz D, Donath C, Sander C (1983) Nature (London) 306:704-707
5. DeFeo-Jones D, Scolnick EM, Koller R, Dhar R (1983) Nature (London) 306:707-709
6. Powers S, Kataoka T, Fasano O, Goldfarb M, Strathern J, Broach J, Wigler M (1984) Cell 36:607-612
7. Toda T, Uno I, Ishikawa T, Powers S, Kataoka T, Broek D, Cameron S, Broach J, Matsumoto K, Wigler M (1985) Cell 40:27-36
8. Uno I, Mitzusawa H, Matsumoto K, Tanaka K, Oshima T, Ishikawa T (1985) Proc Natl Acad Sci USA 82:7855-7859
9. Beckner SK, Hattori S, Shih TY (1985) Nature (London) 317:71-72
10. Birchmeier C, Broek D, Wigler M (1985) Cell 43:615-621
11. Schmitt HD, Wagner P, Pfaff E, Gallwitz D (1986) Cell, in press
12. Murray AW, Szostak JW (1983) Cell 34:961-970
13. Halliday K (1984) J Cyclic Nucleotide Prot Phosphoryl Res 9:435-448
14. Möller W, Amons R (1985) FEBS Lett 186:1-7
15. Tanabe T, Nukada T, Nishikawa Y, Sugimoto K, Suzuki H, Takahashi H, Noda M, Haga T, Ichiyama A, Kangawa K, Minamino N, Matsuo H, Numa S (1985) Nature (London) 315:242-245
16. Jurnak F (1985) Science 230:32-36
17. Fasano O, Aldrich T, Tamanoi F, Taparowsky E, Furth M, Wigler M (1984) Proc Natl Acad Sci USA 81:4008-4012
18. Seeburg PH, Colby WW, Capon JJ, Goeddel DV, Levinson AD (1984) Nature (London) 314:71-74
19. Yuasa Y, Srivastava SK, Dunn CY, Rhim JS, Reddy EP, Aaranson SA (1983) Nature (London) 303:775-779
20. Der CJ, Finkel T, Cooper GM (1986) Cell 44:167-176
21. Shih TY, Stokes PE, Smythers GV, Dhar R, Oroszian S (1982) J Biol Chem 257: 11776-11773
22. McCormick F, Clark BFC, La Cour TFM, Kjelgaard M, Norskov-Lauritsen L, Nijborg J (1985) Science 230:78-82
23. Gibbs JB, Sigal IS, Poe M, Scolnick EM (1984) Proc Natl Adad Sci USA 81:5704-5708
24. Kataoka T, Powers S, Cameron S, Fasano O, Goldfarb M, Broach J, Wigler M (1985) Cell 40:19-26
25. DeFeo-Jones D, Tatchell K, Robinson LC, Sigal IS, Vass WC, Lowy DR, Scolnick EM (1985) Science 228:179-184
26. Fukui Y, Kozasa T, Kaziro Y, Takeda T, Yamamoto M (1986) Cell 44:329-336

Lineage-Specific Oncogene Cooperativity in Avian Hematopoietic Cells

T. Graf, P. Kahn, A. Leutz, B. Vennström, and H. Beug[1]

An increasing body of evidence indicates that multiple events are required for the development of human leukemia. It is likely that the simultaneous or sequential activation of proto-oncogenes plays an important role in this process.

To analyze the effects of oncogene combinations on hematopoietic cells we have utilized an in vitro animal model system consisting of defective leukemia viruses capable of transforming bone marrow cells in culture. In particular, we have studied two chicken retrovirus strains that have acquired two different oncogenes during evolution and that affect the erythroid and myeloid lineages of differentiation, respectively.

Before reviewing our results it is useful to consider how normal hematopoietic cells grow and differentiate. Mature hematopoietic cells are generated from a small population of pluripotent stem cells which have the ability to either self-renew or to become committed to various lineages of differentiation. Once committed, the cells enter an irreversible maturation pathway in which process their proliferative capacity is gradually diminished and eventually lost. The survival, growth, and differentiation of hematopoietic cells is subject to regulation by specific polypeptide growth factors (reviewed in Metcalf, 1984).

Although it is conceivable that oncogenes may affect the self-renewal and/or commitment of pluripotent stem cells, there are as yet no clear examples of this. The better-understood effects of oncogenes occur at the level of the committed progenitors or, within certain lineages, in more mature cells. As outlined in Fig. 1, four basic oncogene effects on the proliferation/differentiation behavior of hematopoietic cells can be distinguished: (1) Induction of proliferation, (2) abolishment of growth factor dependence by an autocrine mechanism, (3) abolishment of growth factor dependence by a non-autocrine (or "bypass") mechanism and (4) block of differentiation. As will be discussed in the following, the four oncogenes contained in the two model retroviruses provide examples for each of these possible effects.

The avian erythroblastosis virus (AEV strain ES4) contains the *erbA* and *erbB* oncogenes and induces acute erythroleukemia as well as sarcomas. The MH2 virus contains the *mil* and *myc* oncogenes and induces monocytic leukemia as well as kidney carcinomas (for review, see Graf and Beug, 1978; Graf and Stehelin, 1982). Both strains transform hematopoietic cells in culture. Cells transformed by AEV resemble erythroblasts and cells transformed by MH2 macrophages, thus corresponding in phenotype to the respective in vivo transformed cells (Gazzolo et al., 1979; Beug et al., 1979). For both virus strains the target cells

[1] Differentiation Programme, European Molecular Biology Laboratory
Meyerhofstrasse 1, 6900 Heidelberg, F.R.G.

37. Colloquium Mosbach 1986
Cell Cycle and Oncogenes

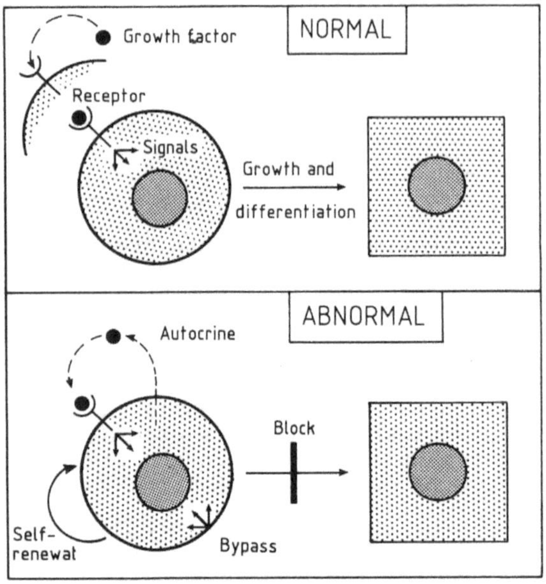

Fig. 1. Schematic representation of the regulation of growth and differentiation in normal hematopoietic cells, and possible alterations induced by oncogenes

are recruited from committed cells: those of AEV correspond to erythroblasts and those of MH2 to myelomonocytic cells (Graf et al., 1981; Samarut and Gazzolo 1982).

To analyze the role of the oncogenes in AEV and MH2, the biological properties of viral mutants deleted in one or the other oncogene were compared to the parental virus. Deletion of the *erbB* oncogene from AEV generates an (*erbA*-containing) virus that lacks detectable biological activity. In contrast, deletion of *erbA* leads to an (*erbB*- containing) virus that is capable of transforming erythroblasts in vivo and in vitro (Frykberg et al., 1983). The *erbB*- transformed erythroblasts are, however, only partly transformed in that they differentiate into mature erythrocytes at a certain frequency (Frykberg et al., 1983; Beug et al. 1985). Both processes, proliferation and differentiation, are independent of the presence of erythropoietin, a growth factor required by normal erythroid progenitors. In addition, the transformed cells do not appear to produce their own growth factor, suggesting a bypass-type mechanism of growth control induced by the oncogene (Beug et al., 1985 and unpublished observations). The effect of the *erbA* oncogenes can be deduced from a more detailed analysis of the phenotype of erythroblasts transformed by both *erbA* and *erbB*. In contrast to *erbB*- only transformed erythroblasts, these cells are completely blocked in their ability to differentiate (Frykberg et al., 1983; Beug et al., 1985; Kahn et al., 1986). The same differentiation-blocking ability of *erbA* could also be shown in combination with other erythroblast-transforming oncogenes such as *src*, *fps*, and Ha-*ras* (Kahn et al., 1984; Kahn et al., 1986). Furthermore, *erbA* is able to convert the sarcoma-inducing potential of the *src* oncogene of Rous sarcoma virus into an acute leukemia-inducing potential, as demonstrated by a viral construct containing both *erbA* and *src* oncogenes (Kahn et al., 1986). These results suggest that the *erbA* gene product does neither specifically nor directly interact with the *erbB* gene product but affects the differentiation capacity of erythroid cells by an independent mechanism.

Deletion of the *myc* gene of MH2 leads to a (*mil*-containing) virus that shows essentially no biological activity. In contrast, deletion of the

Table 1. Cooperativity of retroviral oncogenes in avian hematopoietic cells

Virus strain	"Primary" oncogene	"Auxiliary" oncogene	Target cell lineage
AEV	*erbB*	*erbA*	
	src		
	fps		Erythroid
	H-*ras*		
MH2	*myc*	*mil*	
		erbB	Myeloid
		fps	

mil oncogene leads to a (*myc*- containing) virus that is still capable of efficiently transforming macrophages in vitro but is greatly impaired in its ability to cause monocytic leukemia (Graf et al., 1986). A comparison of macrophages transformed by *myc* alone with those transformed by *mil* plus *myc* shows that in this system the two cell types do not differ in their state of differentiation, but in their dependence on hematopoietic growth factors: while *myc*-transformed cells only proliferate in the presence of cMGF, a chicken growth factor that induces the formation of normal macrophage colonies (Leutz et al., 1984), *mil, myc*-transformed cells proliferate even in its absence (Graf et al., 1986). As shown with antibodies to cMGF, the latter cells secrete cMGF which they themselves require, indicating that *mil* induces autocrine growth of *myc*-transformed myelomonocytic cells (Graf et al., 1986; Weizsäcker et al., 1986). This property is not restricted to *mil*-containing cells since other oncogenes such as *erbB*, *src* and *fps* are similarly capable of inducing cMGF production (Graf et al., 1986).

As summarized in Table 1, our results show that in both model virus systems one of the two oncogenes acts as a "primary oncogene" capable of inducing cell proliferation in specific target cells - the AEV *erbB* gene in erythroblasts, the MH2 *myc* gene in macrophages. The second oncogene present in each virus acts as an "auxiliary oncogene" that enhances the transformed phenotype, either by blocking the differentiation (as is the case for *erbA* in erythroid cells) or by inducing factor production (as is the case for *mil* in myelomonocytic cells). Our results also demonstrate that, depending on the cell type, a given oncogene can exhibit different functions. Thus, *erbB* causes the factor-independent proliferation of erythroid cells while it induces the production of cMGF in *myc*-transformed macrophages without altering their proliferative capacity.

Our studies suggest new possible mechanisms of oncogene cooperativity and have given us insights into the complex mechanisms of hematopoietic cell transformation by viral oncogenes. It remains to be seen whether such mechanisms and/or the particular combinations of oncogenes examined here will also be relevant for other systems, especially for human leukemia.

References

Beug H, v. Kirchbach A, Doederlein G, Conscience J-F, Graf T (1979) Chicken hematopoietic cells transformed by seven strains of defective avian leukemia viruses display three distinct phenotypes of differentiation. Cell 18:375-390

Beug H, Kahn P, Doederlein G, Hayman MJ, Graf T (1985) Characterization of hematopoietic cells transformed in vitro by AEV-H, an *erb*-containing avian eryhtroblastosis virus. In: Neth, Gallo, Greaves, Janka (eds) Modern Trends in Human Leukemia VI, Springer Verlag, Berlin Heidelberg, Vol. 29:290-297

Frykberg L, Palmieri S, Beug H, Graf T, Hayman MJ, Vennstroem B (1983) Transforming capacities of avian erythroblastosis virus mutatns deleted in the *erbA* or *erbB* oncogenes. Cell 32:227-238

Gazzolo L, Moscovici C, Moscovici MG, Samarut J (1979) Response of hemopoietic cells to avian acute leukemia viruses: effects on the differentiation of the target cells. Cell 16:627-638

Graf T, Beug H (1978) Avian leukemia viruses: interaction with their target cells in vivo and in vitro. BBA Revs. Cancer 516:269-299

Graf T, Stehelin D (1982) Avian leukemia viruses. Oncogenes and genome structure. BBA Revs. Cancer 651:245-271

Graf T, v. Kirchbach A, Beug H (1981) Characterization of the hematopoietic target cells of AEV, MC29 and AMV avian leukemia viruses. Exptl. Cell Res 131:331-343

Graf T, v. Weizsäcker F, Grieser S, Coll J, Stehelin D, Patschinsky T, Bister K, Bechade C, Calothy G, Leutz A (1986) V-*mil* induces autocrine growth and enhanced tumorigenicity in v-*myc* transformed avian macrophages. Cell 45:349-356

Kahn P, Adkins B, Beug H, Graf T (1984) *Src*- and *fps*-containing avian sarcoma viruses transform chicken erythroid cells. PNAS USA 81:7122-7126

Kahn P, Frykberg L, Brady C, Stanley IJ, Beug H, Vennström B, Graf T (1986) V-*erbA* cooperates with sarcoma oncogenes in leukemic cell transformation. Cell 45:349-356

Leutz A, Beug H, Graf T (1984) Purification and characterization of cMGF, a novel chicken myelomonocytic growth factor. EMBO J. 3:3191-3197

Metcalf D (1984) The hemopoietic colony stimulates factors. Elsevier Science Publ., Amsterdam, New York, Oxford

Samarut J, Gazzolo L (1982) Target cells infected by avian erythroblastosis virus differentiate and become transformed. Cell 28:921-929

v. Weizsäcker F, Beug H, Graf T (1986) Temperature-sensitive mutants of MH2 avian leukemia virus that map in the v-*mil* and the v-*myc* oncogene, respectively. EMBO J, *5*, 1521-1527

III. Growth Factors: Primary Effects and Relation to Oncogenes

Platelet-Derived Growth Factor: Structure, Function, and Role in Autocrine Stimulation of Growth

C.-H. Heldin[1] and B. Westermark[2]

Introduction

Platelet-derived growth factor is the major mitogen in serum for connective tissue-derived cells and glia cells. Since its discovery more than 10 years ago, considerable interest has been focused on PDGF, initially as a possible causative agent in the development of atherosclerosis (Ross and Glomset 1976), and more recently as a tool in attempts to elucidate the molecular mechanisms of cell transformation (reviewed in Heldin and Westermark 1984).

PDGF (for a review see Heldin et al. 1985) has an M_r of about 30,000 and is composed of two different disulfide-bonded polypeptide chains, denoted A and B. The observation that the mitogenic activity of PDGF is irreversibly lost after reduction, suggests that the dimer form is functionally important; it is not known, however, whether PDGF is a heterodimer or a mixture of homodimers. PDGF interacts with a transmembrane 185 kDa cell surface receptor that is associated with a tyrosine kinase activity in its cytoplasmic domain. Binding of PDGF stimulates the tyrosine kinase activity leading to autophosphorylation of the receptor, as well as to phosphorylation on tyrosine residues of certain cytoplasmic proteins. The significance of tyrosine phosphorylation in the mitogenic pathway remains to be established; such a role, however, is anticipated, since many other growth factor receptors and several oncogene products are kinases with the same unusual amino acid specificity. Other intracellular signals that may be important in the mitogenic pathway include elevation of the cytoplasmic Ca^{2+}-concentration, activation of protein kinase C, cytoplasmatic alkalinization, and induction of specific genes. In the normal cell, this cascade of events is initiated by exogenous PDGF. However, if the mitogenic pathway is shortcircuited at any level, a constitutive growth stimulus may result. We will here review data indicating that the PDGF-dependent signal may be initiated by endogenous production of growth factors which bind to and activate the PDGF receptor.

The Transforming Activity of Simian Sarcoma Virus is Exerted by an Externalized PDGF-Like Growth Factor

Amino acid sequence analysis of PDGF revealed that the A and B chains are homologous to each other. Furthermore, the B chain is almost identical to part of $p28^{sis}$, the transforming protein of simian sarcoma virus (SSV), indicating that SSV has acquired the gene for one of the

[1] Ludwig Institute for Cancer Research (Uppsala Branch), Box 595, BMC, S-751 23 Uppsala, Sweden

[2] Department of Pathology, University Hospital, S-751 85 Uppsala, Sweden

37. Colloquium Mosbach 1986
Cell Cycle and Oncogenes

two polypeptide chains of PDGF (Waterfield et al. 1983, Doolittle et al. 1983). This conclusion has been verified by cloning of c-*sis*, the cellular counterpart to v-*sis* (Josephs et al. 1984a, A. Johnsson et al. 1984, Chiu et al. 1984). The human c-*sis* transcript has a size of 3.5-4.2 kb and contains an open reading frame for a 28 kDa PDGF B chain precursor, but does not code for the A chain (Collins et al. 1985).

The finding of a structural homology between PDGF and a transforming protein led to the hypothesis that the transforming activity of SSV is exerted by a PDGF-like growth factor. The target cell specificity of SSV is in support of this hypothesis; SSV transforms cells that carry PDGF receptors (Deinhardt 1980). SSV-transformed cells have also been found to produce a PDGF-like growth factor (Deuel et al. 1983, Bowen-Pope et al. 1984, Owen et al. 1984, Garrett et al. 1984, Huang et al. 1984, A. Johnsson et al. 1985a, Leal et al. 1985), which in its mature from has an M_r of 24,000 and appears to be structurally similar to a PDGF B chain homodimer (Robbins et al. 1983). Furthermore, the growth and transformed morphology of SSV-transformed human fibroblasts are reverted by exogenously added PDGF-antibodies (A. Johnsson et al. 1985b).

The ability of PDGF antibodies to inhibit SSV-transformation indicates that the transforming product is externalized. Evidence that externalization of the v-*sis* product is in fact necessary for its function, was obtained by experiments where the v-*sis* gene was systematically modified. The transforming activity was lost when the 5' flanking *env*-derived sequence was deleted (Hannink and Donoghue 1984, King et al. 1985). This sequence codes for a hydrophobic stretch of amino acids which probably serves as a signal sequence and allows the transport of the product into secretory vesicles. Interestingly, c-*sis* has also a 5' part coding for a signal sequence (Jospehs et al. 1984b), that is required for its transforming activity when the gene is transfected into NIH 3T3 cells (Gazit et al. 1984).

The similarities in functional activities between the v-*sis* product and PDGF, and the similarities in morphology and growth behavior of SSV-transformed cells and PDGF-stimulated cells, indicate that the v-*sis* product is a full PDGF agonist, but has no additional cellular effects. Thus, SSV most likely provides only an efficient growth stimulus to infected cells, without qualitative differences compared with those obtained via exogenously added PDGF. The acquisition of the fully malignant phenotype involving also immortalization, capacity for invasiveness, and formation of metastases, clearly require additional genetic alterations. Notably, however, SSV induces malignant gliomas in newborn marmosets (L. Johnsson et al. 1975). It is possible that SSV has the capacity to induce a fully malignant phenotype in certain vulnerable cells present in the neonatal brain. Alternatively, a constitutive growth signal may significantly increase the risk for additional genetic changes contributing to malignant transformation at a certain stage of differentiation.

Cell Lines of Human Tumors Produce PDGF-Like Growth Factors

Several years ago the human osteosarcoma cell line, U-2 OS, was shown to produce a growth factor which was found to be structurally, functionally and immunologically related to PDGF (Heldin et al. 1980). Since this cell type is assumed to respond to PDGF, it was hypothesized that the endogenously produced growth factor was of importance in

Table 1. Production of PDGF-like growth factors by various cell types

Nontransformed cells	
Smooth muscle cells	Seifert et al. 1984, Nilsson et al. 1985
Cytotrophoblasts	Goustin et al. 1985
Endothelial cells	DiCorleto and Bowen-Pope 1983
Macrophages	Shimokado et al. 1985, Martinet et al. 1985
Established cell lines	
Human	
Osteosarcoma (U-2OS, U-393OS, SAOS-2)	Heldin et al. 1980, Betsholtz et al. 1986
Synovial sarcoma (U-4 SS)	Betsholtz et al. 1986
Giant cell sarcoma (B-5 GT)	Betsholtz et al. 1986
Leiomyosarcoma (SKLMS)	Betsholtz et al. 1986
Glioma (U-343 MGa, U-178 MG)	Nistér et al. 1984, unpublished
Rhabdomyosarcoma (RD)	Betsholtz et al. 1983
Lung carcinoma (U-1810)	Betsholtz et al., unpublished
Breast carcinoma (T47, MCF-7)	Rozengurt et al. 1985, Peres et al., unpublished
Bladder carcinoma (T24)	Bowen-Pope et al. 1984
Hepatoma (HepG2)	Bowen-Pope et al. 1986
Melonoma (WM 115)	Westermark et al. 1986
Nonhuman	
Myogenic cell line (L6J1)	Sejersen et al., unpublished
Teratocarcinoma (PSA-1-G)	Gudas et al. 1984
Neuroblastoma (Neuro 2A)	Van Zoelen et al. 1985
Virus-transformed cells	
SSV-transformed cells	Deuel et al. 1983
Kirsten murine sarcoma virus-transformed cells	Bowen-Pope et al. 1984
Moloney murine sarcoma virus-transformed cells	Bowen-Pope et al. 1984
Adenovirus-transformed cells	Bowen-Pope et al. 1984
SV-40-transformed cells	Stroobant et al. 1985, Betsholtz et al. 1985

Cell types are lsited that have been shown to produce PDGF receptor competing activity or a nonreduced 31 kDa component that is recognized by PDGF antibodies

autocrine stimulation of cell growth. Recently, evidence was obtained that autocrine receptor activation occurs in this cell line (Betsholtz et al. 1984). Screening, by use of receptor competition assays, has now revealed that cell lines established from tumors of various histogenetic origins secrete PDGF-like growth factors (Table 1). Notably, not only PDGF-responsive cell types are represented, but also, for example, hepatoma cells, lung carcinoma cells, breast carcinoma cells, and neuroblastoma cells. Since these cell types lack PDGF receptors, it is unlikely that the endogenous production of a growth factor has any impact on the growth of the cells. It is, however, an interesting possibility that this reflects the production of PDGF-like factors by tumor cells in vivo, which may stimulate growth of surrounding connective tissue cells in a paracrine manner, and cause the commonly found stromal infiltration and proliferation in tumors.

Immunoprecipitation using a PDGF antiserum revealed that the tumor cell-derived factors have M_r:s of 31,000 and are converted to a double band of M_r 17,000 to 16,500 after reduction (Betsholtz et al. 1983).

The osteosarcoma-derived growth factor was recently purified to homogeneity (Heldin et al. 1986). Structural analysis of the purified product revealed, rather unexpectedly, that it is a homodimer of PDGF A chains. Thus, homodimers of either of the two subunits of PDGF bind to and activate the PDGF receptor.

Analysis of a large number of different clones from a human glioma cell line, U-343 Mga, that produces a PDGF-like growth factor (Nistér et al. 1984), revealed a dramatic variability both in the secretion of the PDGF-like growth factor, and in the expression of PDGF receptors. A correlation was found between the growth rate under serum free conditions and the secretion of PDGF-like growth factor, providing indirect evidence for autocrine stimulation of growth (Nistér et al. 1986).

The glioma cell clone that produces the largest quantity of PDGF-like growth factor was used as source of mRNA for the cloning of cDNA for the A chain (Betsholtz et al. 1986). The nucleotide sequence predicts a 211-amino-acid-long A chain precursor, which shows extensive homology to the B chain precursor. Both precursors have a hydrofobic leader sequence indicating that they are secreted. Proteolytic processing occur in the N-terminal parts, after stretches of four and two basic amino acids, respectively. The B chain, and possibly also the A chain, is further processed in the C-terminus, yielding mature polypeptide chains of slightly more than 100 amino acids each.

The A chain probe hybridizes to three major transcripts in U-343 MGa C1 2 cells of sizes 1.9, 2.3 and 2.8 kb, possibly reflecting that differential splicing occurs. The transcripts of the two polypeptide chains of PDGF were found to be frequently and independently expressed in human tumor cell lines; examples were found of cell lines that expressed both transcripts, as well as only either one (Betsholtz et al. 1986). The secretion of a 31 kDa PDGF-like growth factor to the medium was found to correlate exactly to the expression of the A chain transcript, indicating that all such secreted PDGF-like factors may be A chain homodimers. It is not known why B chains have not been observed in the cell culture media of cells expressing the B chain transcript. In analogy with the behavior of the product of SSV-transformed 3T3 cells (Robbins et al. 1985), it is possible that B chains, although externalized from the cells, remain associated with the outer cell surface, or are rapidly cleared from the culture medium.

PDGF-Like Growth Factors from Normal Cells

Cytotrophoblasts, smooth muscle cells, endothelial cells and activated macrophages have recently been found to secrete PDGF-like growth factors under certain conditions, indicating that autocrine and paracrine mechanisms may operate not only in transformed cells, but also in certain normal cells.

The finding that human placental cytotrophoblasts both produce and respond to PDGF-like growth factors provided a clue to the in vivo function of PDGF - to stimulate placental growth under a defined phase of the development (Goustin et al. 1985). By use of in situ hybridization technique the c-*sis* expression was found to correlate with c-*myc* expression, suggesting that the c-*sis* product induced c-*myc* expression (Goustin et al. 1985), analogous to how PDGF induces c-*myc* in 3T3 cells (Kelly et al. 1983).

Smooth muscle cells have also been found to produce PDGF-like growth factors under certain conditions (Seifert et al. 1984, Nilsson et al. 1985). Since smooth muscle cells respond to PDGF, it is possible that an endogenous production of PDGF-like growth factors causes an autocrine loop which stimulates cell growth.

Since endothelial cells and macrophages lack PDGF receptors, it is unlikely that their production of PDGF-like growth factors (DiCorleto and Bowen-Pope 1983, Shimokado et al. 1985, Martinet et al. 1985) serves an autocrine function. Rather it may cause paracrine stimulation of growth of surrounding connective tissue cells, seen, for example, in the vessel wall as a response to injury, or in chronic inflammatory processes.

Evolution of Two Different but Homologous PDGF Subunits

PDGF-like growth factors have been found to be produced by a large variety of different transformed cell types: SSV-transformed cells produce a factor structurally related to a PDGF B chain homodimer, where-

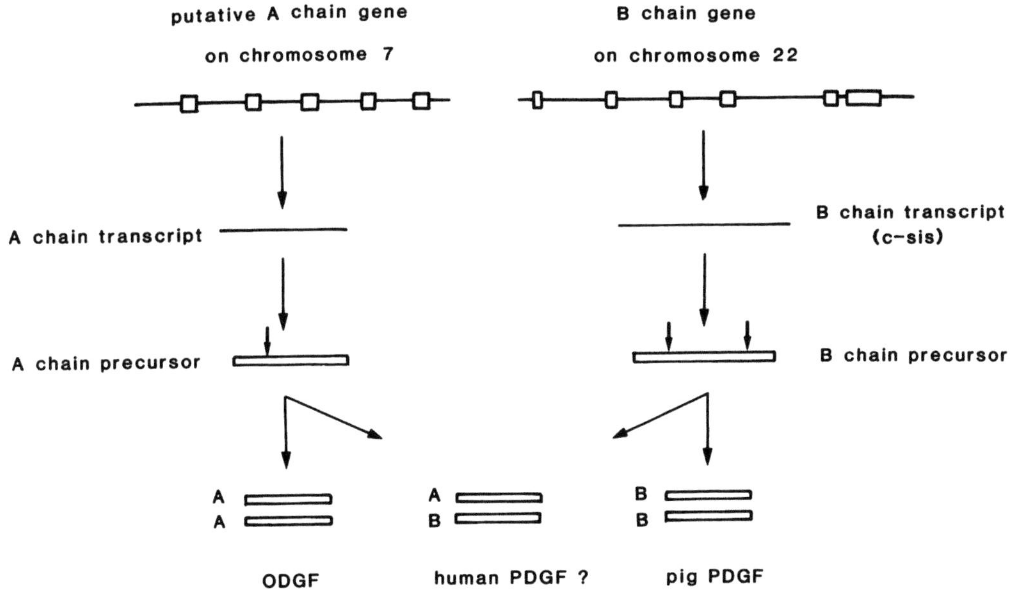

Fig. 1. Schematic illustration of the two genes of PDGF and their products. Heterodimers or homodimers of the two polypeptide chains are formed before or after processing of the protein precursors (indicated by arrows). ODGF and pig PDGF are examples of homodimers of the A and B chains, respectively. Moreover, the product of SSV-transformed cells has a structure similar to a homodimer of PDGF B chains (Robbins et al. 1983). The structure of human PDGF from platelets has not been unequivocally determined; as discussed in the text, indirect evidence indicate that it might be a heterodimer

as several cell lines of human tumors secrete A chain homodimers. Interestingly, pig PDGF appears to be a homodimer of B chains (Stroobant and Waterfield 1984). In spite of the numerous examples of PDGF-like growth factors that have the structure of homodimers, it is possible that human PDGF purified from platelets is a heterodimer; purified fractions of PDGF contain stoichiometric amounts of A and B chains, and it has not been possible to separate homodimers from purified PDGF (unpublished observations) (Fig. 1). It is not clear why two different PDGF chains have evolved. Their sequence homolgy indicates that they originated from a single ancestral gene. After duplication, the two genes acquired different chromosomal locations; the A chain gene is located on chromosome 7 (Betsholtz et al. 1986) and the B chain gene on chromosome 22 (Swan et al. 1982, Dalla Favera et al. 1982). Some evolutionary divergence has occurred, but apparently the receptor binding function has been conserved. As discussed above, the two polypeptide chains show differences in their secretory behavior. It is an interesting possibility that they have also other, yet unraveled, functional differences.

References

Betsholtz C, Heldin C-H, Nistér M, Ek B, Wasteson Å, Westermark B (1983) Synthesis of a PDGF-like growth factor in human glioma and sarcoma cells suggests the expression of the cellular homologue of the transforming protein of simian sarcoma virus. Biochem Biophys Res Commun 117:176-182

Betsholtz C, Westermark B, Ek B, Heldin C-H (1984) Co-expression of a PDGF-like growth factor and PDGF receptors in a human osteosarcoma cell line: implications for autocrine receptor activation. Cell 39:447-457

Betsholtz C, Bywater M, Westermark B, Bürk R, Heldin C-H (1985) Expression of the *c-sis* gene and secretion of a platelet-derived growth factor-like protein by simian virus 40 transformed BHK cells. Biochim Biophys Res Commun 130:753-760

Betsholtz C, Johnsson A, Heldin C-H, Westermark B, Lind P, Urdea MS, Eddy R, Philpott K, Mellor A, Knott TJ, Scott J (1986) The human platelet-derived growth factor A chain: complementary DNA sequence, chromosomal localization and expression in human tumor cell lines. Nature (London)

Bowen-Pope DF, Vogel A, Ross R (1984) Production of platelet-derived growth factor-like molecules and reduced expression of platelet-derived growth factor receptors accompany transformation by a wide spectrum of agents. Proc Natl Acad Sci USA 81:2396-2400

Chiu I-M, Reddy EP, Givol D, Robbins KC, Tronick SR, Aaronson SA (1984) Nucleotide sequence analysis identifies the human c-*sis* proto-oncogene as a structural gene for platelet-derived growth factor. Cell 37:123-129

Collins T, Ginsburg D, Boss JM, Orkin SH, Pober JS (1985) Cultured human endothelial cells express platelet-derived growth factor B chains: cDNA cloning and structural analysis. Nature (London) 316:748-750

Dalla Favera R, Gallo RC, Giallongo A, Croce CM (1982) Chromosomal localization of the human homolog (c-*sis*) of the simian sarcoma virus onc gene. Science 218:686-688

Deinhardt F (1980) Biology of primate retrovirus. In: Klein G (ed) Viral oncology. Raven Press, New York, pp 357-398

Deuel TF, Huang JS, Huang SS, Stroobant P, Waterfield M (1983) Expression of a platelet-derived growth factor-like protein in simian sarcoma virus transformed cells. Science 221:1348-150

DiCorleto PE, Bown-Pope DF (1983) Cultured endothelial cells produce a platelet-derived growth factor-like protein. Proc Natl Acad Sci USA 80:1919-1923

Doolittle RF, Hunkapiller MW, Hood LE, Devare SG, Robbins KC, Aaronson SA, Antoniades HN (1983) Simian sarcoma virus onc-gene, v-*sis*, is derived from the gene (or genes) encoding a platelet-derived growth factor. Science 221:275-277

Garrett JS, Coughlin SR, Niman HL, Tremble PM, Giels GM, Williams LT (1984) Blockade of autocrine stimulation in simian sarcoma virus-transformed cells reverses down-regulation of platelet-derived growth factor receptors. Proc Natl Acad Sci USA 81:7466-7470

Gazit A, Igarashi H, Chiu I-M, Srinivasan A, Yaniv A, Tronick SR, Robbins KC, Aaronson SA (1984) Expression of the normal human sis/PDGF-2 coding sequence induces cellular transformation. Cell 39:80-97

Goustin AS, Betsholtz C, Pfeiffer-Ohlsson S, Persson H, Rydnert J, Bywater M, Holmgren G, Heldin C-H, Westermark B, Ohlsson R (1985) Co-expression of the sis and myc protooncogenes in human placenta suggests autocrine control of trophoblast growth. Cell 41:301-312

Gudas LJ, Singh JP, Stiles CD (1982) Secretion of growth regulating molecules by teratocarcinoma stem cells. In: Silver, Martin, Strickland (eds) Teratocarcinoma stem cells. Cold Spring Harbor Conf Cell Prol 10. Cold Spring Harbor Press, New York, pp 229-236

Hannink M, Donoghue DJ (1984) Requirement for a signal sequence in biological expression of the v-*sis* oncogene. Science 226:1197-1199

Heldin C-H, Westermark B (1984) Growth factors: Mechanism of action and relation to oncogenes. Cell 37:9-20

Heldin C-H, Westermark B, Wasteson Å (1980) Chemical and biological properties of a growth factor from human cultured osteosarcoma cells: resemblance with platelet-derived growth factor. J Cell Physiol 105:235-246

Heldin C-H, Wasteson Å, Westermark B (1985) Platelet-derived growth-factor. Mol Cell Endocrinol 39:169-187

Heldin C-H, Johnsson A, Wennergren S, Wernstedt C, Betsholtz C, Westermark B (1986) A human osteosarcoma cell line secretes a growth factor structurally related to a homodimer of PDGF A chains. Nature (London) 319:511-514

Huang JS, Huang SS, Deuel TF (1984) Transforming proteins of simian sarcoma virus stimulates autocrine growth of SSV-transformed cells through PDGF cell-surface receptors. Cell 39:79-87

Johnsson A, Heldin C-H, Wasteson Å, Westermark B, Deuel TF, Huang JS, Seeburg PH, Gray E, Ullrich A, Scrace G, Stroobant P, Waterfield MD (1984) The c-*sis* gene encodes a precursor of the B chain of platelet-derived growth factor. EMBO J 3: 921-928

Johnsson A, Betsholtz C, Helm K von der, Heldin C-H, Westermark B (1985a) Platelet-derived growth factor agonist acitivity of a secreted form of the v-*sis* oncogene product. Proc Natl Acad Sci USA 81:1721-1725

Johnsson A, Betsholtz C, Heldin C-H, Westermark B (1985b) Antibodies to platelet-derived growth factor inhibit acute transformation by simian sarcoma virus. Nature (London) 317:438-440

Johnsson L Jr, Wolfe LG, Whisler WW, Norton T, Thakkar B, Deinhardt F (1975) Induction of gliomas in marmosets by simian sarcoma virus, typ 1 (SSV-1). Abstr Proc Am Assoc Cancer Res 16:119

Josephs SF, Guo C, Ratner L, Wong-Staal F (1984a) Human proto-oncogene nucleotide sequences corresponding to the transforming region of simian sarcoma virus. Science 223:487-491

Josephs SF, Ratner L, Clarke MF, Westin EH, Reitz MS, Wong-Staal F (1984b) Transforming potential of human c-*sis* nucleotide sequences encoding platelet-derived growth factor. Science 225:636-639

Kelly K, Cochran BH, Stiles CD, Leder P (1983) Cell-specific regulation of the c-*myc* gene by lymphocyte mitogens and platelet-derived growth factor. Cell 35:603-610

King CR, Giese NA, Robbins KC, Aaronson SA (1985) In vitro mutagenesis of the v-*sis* transforming gene defines functional domains of its growth factor-related product. Proc Natl Acad Sci USA 82:5295-5299

Leal F, Williams LT, Robbins KC, Aaronson SA (1985) Evidence that the v-*sis* product transforms by interaction with the receptor for platelet-derived growth factor. Science 230:327-330

Martinet Y, Bitterman PB, Mornex J-F, Grotendorst G, Martin GR, Crystal RG (1985) Activated human monocytes express the c-*sis* proto-oncogene and release a mediator showing PDGF-like activity. Nature (London) 319:158-160

Nilsson J, Sjölund M, Palmberg L, Thyberg J, Heldin C-H (1985) Arterial smooth muscle cells in primary culture produce a platelet-derived growth factor-like protein. Proc Natl Acad Sci USA 82:4418-4422
Nistér M, Heldin C-H, Wasteson Å, Westermark B (1984) A glioma-derived analog to platelet-derived growth factor: demonstration of receptor competing activity and immunological cross-reactivity. Proc Natl Acad Sci USA 81:926-930
Nistér M, Heldin C-H, Westermark B (1986) Clonal variation in the production of a PDGF-like growth factor and expression of PDGF receptors in a human malignant glioma. Cancer Res 46:332-340
Owen AJ, Pantazis P, Antoniades HN (1984) Simian sarcoma virus-transformed cells secrete a mitogen identical to platelet-derived growth factor. Science 225:54-56
Robbins KC, Antoniades HN, Devare SG, Hunkapiller MW, Aaronson SA (1983) Structural and immunological similarities between simian sarcoma virus gene product(s) and human platelet-derived growth factor. Nature (London) 305:605-608
Robbins KC, Leal F, Pierce JH, Aaronson SA (1985) The v-sis/PDGF-2 transforming gene product localizes to cell membranes but is not a secretory protein. EMBO J 4: 1783-1792
Ross R, Glomset JA (1976) The pathogenesis of artherosclerosis. N Engl J Med 295: 369-377
Rozengurt E, Sinnett-Smith J, Taylor-Papadimitriou J (1985) Production of PDGF-like growth factor by breast cancer cell lines. Int J Cancer 36:247-252
Seifert RA, Schwartz SM, Bowen-Pope DF (1984) Developmentally regulation of production of platelet-derived growth factor-like molecules. Nature (London) 311:669-671
Shimokado K, Raines ER, Madtes DK, Barrett TB, Benditt E, Ross R (1985) A significant part of macrophage-derived growth factor consists of at least two forms of PDGF. Cell 43:277-286
Stroobant P, Waterfield MD (1984) Purification and properties of porcine platelet-derived growth factor. EMBO J 2:2963-2967
Stroobant P, Gullick WJ, Waterfield MD, Rozengurt E (1985) Highly purified fibroblast-derived growth factor, an SV 40-transformed fibroblast mitogen, is closely related to platelet-derived growth factor. EMBO J 4:1945-1949
Swan DC, McBride OW, Robbins KC, Keithly DA, Reddy EP, Aaronson SA (1982) Chromosomal mapping of the simian sarcoma virus onc gene analogue in human cells. Proc Natl Acad Sci USA 79:4691-4695
Waterfield MD, Scrace G, Whittle N, Stroobant P, Johnsson A, Wasteson Å, Westermark B, Heldin C-H, Huang JS, Deuel TF (1983) Platelet-derived growth factor is structurally related to the putative transforming protein of $p28^{sis}$ of simian sarcoma virus. Nature (London) 304:35-39
Westermark B, Johnsson A, Paulsson Y, Betsholtz C, Heldin C-H, Herlyn M, Radeck U, Koprowski H (1986) Human melanoma cell lines of primary and metastatic origin express the genes encoding the constituent chains of PDGF and produce a PDGF-like growth factor. Proc Natl Acad Sci USA
Zoelen EJJ van, Ven WJM van de, Franssen HJ, Oostwaard TMJ van, Saag PT van der, Heldin C-H, Laat SW de (1985) Neuroblastoma cells express c-*sis* and produce a transforming growth factor antigenically related to the platelet-derived growth factor. Mol Cell Biol 5:2289-2297

Signal Transduction by Growth Factor Receptors

W. H. Moolenaar, A. J. Bierman, L. H. K. Defize, B. C. Tilly, and S. W. de Laat[1]

Introduction

Growth factors are polypeptide hormones that induce replicative DNA synthesis and cell division in their target cells. Like all polypeptide hormones, growth factors initiate their action by binding to specific, high-affinity receptor molecules on the cell surface. Although much has been learned about growth factor-receptor interaction, the molecular mechanisms of action and the in vivo function of growth factors are poorly understood.

Following growth factor binding, the activated receptor mediates a cascade of rapid biochemical and physiological changes in the cell, which ultimately (after 10-20 h) lead to the stimulation of DNA synthesis. As a rule, the growth factor has to be present throughout the entire "pre-replicative" phase (usually 8-10 h) for commitment to DNA synthesis to occur. It turns out that there are at least three potential signal pathways in the action of growth factors such as EGF and PDGF (Fig. 1).

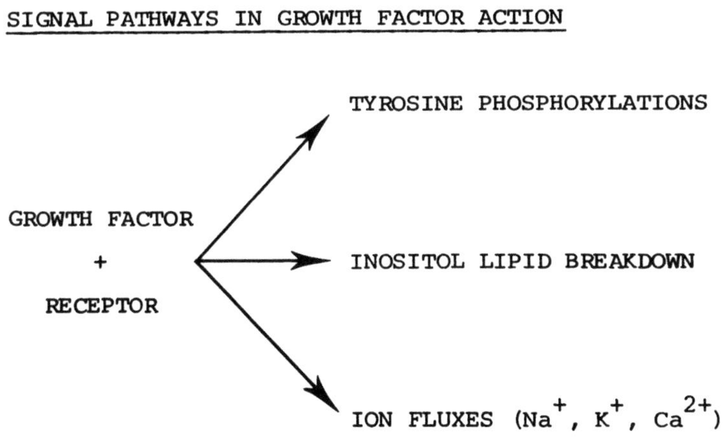

Fig. 1. Potential signaling pathways in the action of growth factors

One of the first consequences of growth factor-receptor interaction is the activation of a protein kinase specific for tyrosyl residues (Carpenter 1984, Heldin and Westermark 1984, Hunter and Cooper 1985). In fact, the receptors for growth factors like EGF and PDGF are transmembrane glycoproteins that possess intrinsic tyrosine-specific protein kinase activity (Hunter and Cooper 1985). Growth factor binding indu-

[1]Hubrecht Laboratory, Uppsalalaan 8, 3584 CT Utrecht, The Netherlands

37. Colloquium Mosbach 1986
Cell Cycle and Oncogenes

ces a rapid stimulation of the receptor kinase, resulting in autophosphorylation of the receptor itself, as well as in the phosphorylation of various substrate proteins. The intrinsic tyrosine-specific kinase activity is shared with several viral oncogene products, such as the transforming protein of Rous sarcoma virus (Bishop 1985). This suggests that tyrosine-specific protein phosphorylations may initiate a set of common mitogenic pathways in virus-transformed and growth factor-stimulated cells. However, it is not yet possible to relate increased tyrosine kinase activity to specific metabolic alterations in stimulated cells.

Other immediate consequences of receptor activation (Fig. 1) include the breakdown of inositol phospholipids (Sawyer and Cohen 1981, Berridge 1984), a transient rise in cytoplasmic free Ca^{2+} ($[Ca^{2+}]_i$) (Moolenaar et al. 1984, 1986), and the stimulation of monovalent ion transport across the plasma membrane (Rozengurt 1981, Moolenaar 1986a). Of the known ionic transport changes in growth factor-stimulated cells, the activation of electroneutral Na^+/H^+ exchange is best characterized (see next section).

Activation of Na^+/H^+ Exchange

The first direct evidence that growth factors activate an otherwise quiescent Na^+/H^+ exchanger in the plasma membrane came from studies on serum-stimulated neuroblastoma cells (Moolenaar et al. 1981a,b). Re-addition of serum to growth-arrested cells leads to the rapid activation of an electrically silent Na^+ influx pathway, which is sensitive to amiloride and can be stimulated by acidifying the cytoplasm using weak acids. In addition, it was shown that amiloride-sensitive Na^+ uptake in such acid-loaded cells is coupled to the efflux of H^+ with an 1:1 stoichiometry. In a subsequent study it was shown that EGF-induced Na^+ influx in quiescent human fibroblasts is amiloride-sensitive, electroneutral, and is enhanced by cytoplasmic acid loads (Moolenaar et al. 1982). From these studies it was predicted that stimulation by mitogens leads to an increase in pH_i, while the accompanying entry of Na^+ results in a stimulation of the Na^+, K^+-pump. New developments in pH_i-monitoring techniques, particularly the synthesis of fluorescent indicators that can be trapped into the cytoplasm of small cells, have allowed to show that the Na^+/H^+ exchanger is normally involved in the close regulation of pH_i and that activation of the Na^+/H^+ exchanger by growth factors leads to a sustained rise in pH_i of 0.2-0.3 unit.

Most cells maintain their pH_i at 7.0-7.4. This is well above the electrochemical equilibrium value of 6.0-6.4 that is predicted by the Nernst equation from a transmembrane potential of approximately -60 mV. In vertebrate cells, the specific H^+-extruding mechanism which raises pH_i appears to be Na^+/H^+ exchange (Roos and Boron 1981, Moolenaar 1986b). The functioning of the Na^+/H^+ exchanger in the plasma membrane and its role in pH_i homeostasis is most easily assessed by continuously monitoring the rapid recovery of pH_i to its resting level after a sudden acidification of the cytoplasm, as induced by a NH_4^+-prepulse or by weak acids (Roos and Boron 1981). In most cells, this pH_i recovery process follows an exponential time course and is entirely due to net H^+ extrusion via the Na^+/H^+ exchanger, which utilizes the energy stored in the transmembrane Na^+ gradient. The major determinant of the rate of the Na^+/H^+ exchanger is pH_i. At normal pH_i values (near 7.0), the exchanger is relatively inactive, although the steep transmembrane Na^+ gradient could theoretically raise pH_i about one unit more alkaline.

As pH_i falls below a certain "threshold", the Na^+/H^+ exchanger is increasingly stimulated. Aronson and co-workers (Aronson et al. 1982) were the first to point out that the Na^+/H^+ exchanger is apparently set in motion through allosteric activation by cytoplasmic H^+ at a regulatory site which is distinct from the internal H^+ transport binding site. The relatively strong pH_i sensitivity of the exchanger is, of course, a crucial property for an H^+ extruding system to maintain pH_i at a critical level. Furthermore, a change in the apparent affinity of the exchanger for internal H^+ could provide a powerful mechanism by which external stimuli regulate the physiological state of the exchanger.

The diuretic amiloride reversibly inhibits Na^+/H^+ exchange, apparently by competing with Na^+ for binding to the same external site (K_i 2-5 μM). Several amiloride analogs have been found to be 10-100 times more potent than amiloride in blocking Na^+/H^+ exchange (l'Allemain et al. 1984, Zhuang et al. 1984). It should be noted, however, that amiloride and its potent analogs are taken up by most cells in culture and thereby may affect cellular functions such as protein synthesis and kinase activity (Zhuang et al. 1984, Besterman et al. 1985, Davis and Czech 1985).

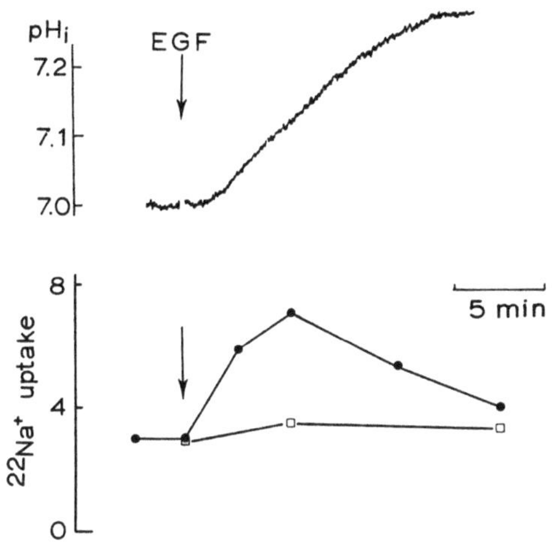

Fig. 2. Activation of Na^+/H^+ exchange by EGF in human fibroblasts. *Upper panel* shift in pH_i induced by EGF (10 ng/ml). Quiescent monolayers were loaded with BCECF as described (Moolenaar et al. 1983). *Lower panel* time course of EGF-induced $^{22}Na^+$ uptake in the presence (●) and absence (□) of 1 mM amiloride. Initial rates of $^{22}Na^+$ uptake were measured over 3 min intervals in the presence of 0.1 mM ouabain to block active Na^+ efflux

Figure 2 shows a typical example of an alkaline pH_i shift after addition of EGF to quiescent cells loaded with the pH-sensitive dye bis (carboxyethyl)carboxyfluorescein (BCECF). The shift in pH_i is initiated within 20-30 sec and is complete by 10-15 min. The elevated pH_i persists for as long as the growth factor is present. In general, the induced alkalinizations range from 0.1-0.3 pH unit; they are inhibited by amiloride and by Na^+ removal and are accompanied by a transient increase in amiloride-sensitive $^{22}Na^+$ uptake (Fig. 1; for review see Moolenaar 1986a). Furthermore, the rise in pH_i is converted into a fall in pH_i when the direction of the transmembrane Na^+ gradient is reversed (Moolenaar et al. 1983). When taken together, these data convincingly demonstrate that the mitogen-induced pH_i rise is mediated by the Na^+/H^+ exchanger.

Mechanism of Activation of Na^+/H^+ Exchange

How does receptor occupancy lead to activation of the Na^+/H^+ exchanger? Recent studies have shown that the activation is attributable to an alkaline shift in the pH_i sensitivity of the exchanger (Moolenaar et al. 1983, Paris and Poysségur 1984, Grinstein et al. 1985). As mentioned above, this pH_i sensitivity is determined by an allosteric H^+-binding site on the cytoplasmic face of the exchanger. It thus seems plausible to assume that the altered pH_i sensitivity of the exchanger is due to some conformational change resulting in an increased pK_a of the regulatory H^+-binding site. Thus, the physiological effect of growth factors on the Na^+/H^+ exchanger is to increase its pH_i threshold, that is the level to which pH_i must rise before the exchanger virtually shuts off. Indeed, the Na^+/H^+ exchanger is only transiently stimulated by external stimuli and its activity returns to the control level once pH_i has attained its new stable value (Fig. 2 and Moolenaar 1986a).

The obvious next question is by which route growth factors modify the pH_i sensitivity of the exchanger to raise pH_i. An important key comes from studies using the tumour promotor 12-O-tetradecanoyl-13-acetate (TPA), which binds to and directly activates the phospholipid-dependent protein kinase C (Nishizuka 1984). Under normal conditions, kinase C is activated by endogenous diacylglycerol (DG) derived from the breakdown of inositol phospholipids. Both TPA and synthetic DG are capable of mimicking the effects of growth factors on Na^+/H^+ exchange and pH_i in various cell types (reviewed by Moolenaar 1986a). The simplest explanation for those findings is that kinase C directly phosphorylates the exchanger, but it cannot, of course, be excluded that kinase C acts in a more indirect way to activate Na^+/H^+ exchange. Information concerning the molecular structure of the exchange carrier may become available in the near future, and this should greatly facilitate further study of the role of kinase C and other kinases in the activation process.

The apparent involvement of kinase C strongly suggests that an alkaline shift in pH_i is not uniquely induced by growth factors and phorbol esters but may be a common cellular response in the action of those hormones and neurotransmitters that trigger the hydrolysis of inositol phospholipids and thereby generate diacylglycerol. There is some evidence to support this notion: in a typical model systems of hormone-induced phosphoinositide turnover, such as neutrophils, the pH rises rapidly after activation by physiological stimuli (Molski et al. 1980). Finally, it is noteworthy that there is evidence, albeit circumstantial, that pathways other than kinase C may be involved in Na^+/H^+ exchange activation (Vara and Rozengurt 1985, Paris and Poysségur 1986). The biochemical nature of these pathways remains to be elucidated.

Ca^{2+} Mobilization by Growth Factors

In addition to activating Na^+/H^+ exchange, mitogens like EGF and PDGF induce a rapid but transient rise in $[Ca^{2+}]_i$ in their target cells, as measured by quin-2 fluorescence (Moolenaar et al. 1983, 1986, Hesketh et al. 1985). Addition of EGF or PDGF to responsive cells elicits a severalfold rise in $[Ca^{2+}]_i$ which is initiated without a detectable lag period and is usually complete within 30-60 s. Thereafter, $[Ca^{2+}]_i$ gradually returns to near-basal levels over a 10-min period. Figure 3 schematically illustrates the time courses of both the $[Ca^{2+}]_i$ transient and the rise in pH_i as induced by EGF in responsive cells.

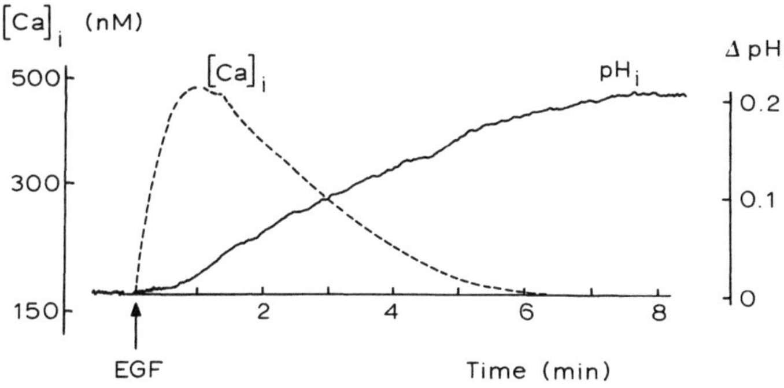

Fig. 3. Changes in $[Ca]_i$ and pH_i in human A431 cells following the addition of EGF (100 ng/ml). For methods see Moolenaar et al. (1983, 1986)

How does growth factor binding increase $[Ca^{2+}]_i$? The Ca^{2+} signal in response to PDGF is not prevented by removal of external Ca^{2+}, indicating that the Ca^{2+} is released from intracellular stores, presumably the endoplasmic reticulum (Moolenaar et al. 1984). Indeed, PDGF provokes the rapid formation of inositol-1,4,5-trisphosphate [IP_3(1,4,5)], the key messenger for mobilizing Ca^{2+} from nonmitochondrial stores (Berridge et al. 1984, Berridge and Irvine 1984). Surprisingly, the Ca^{2+} signal in response to EGF shows no contribution from intracellular stores, but seems to result from net Ca^{2+} entry through a voltage-independent Ca^{2+} channel in the plasma membrane (Moolenaar et al. 1986). This interpretation is based on the finding that the EGF-induced $[Ca^{2+}]_i$ rise in human A431 carcinoma cells is critically dependent on the extracellular Ca^{2+} concentration, is accompanied by enhanced $^{45}Ca^{2+}$ uptake (Sawyer and Cohen 1981), is blocked by Ca^{2+} entry blockers like La^{3+} and Mn^{2+}, while it is not accompanied by changes in transmembrane potential. Since EGF is known to stimulate phosphoinositide turnover (see next section) and diacylglycerol formation in A431 cells (Sawyer and Cohen 1981, Sahai et al. 1982), it is intriguing that EGF does not seem to act on intracellular stores to raise $[Ca^{2+}]_i$.

What could be the molecular nature of the putative Ca^{2+} channel that is activated when EGF binds to its receptor? An attractive candidate for mediating Ca^{2+} entry is phosphatidic acid, an anionic phospholipid that is rapidly formed from diacyl-glycerol in EGF-treated A431 cells (Sawyer and Cohen 1981) and that may function as a Ca^{2+} ionophore in model sstems (Serhan et al. 1982) and in intact cells from various tissues (Putney et al. 1980, Ohsako and Deguchi 1981). Our preliminary experiments (unpublished) indeed indicate that exogenous phosphatidate can transiently elevate $[Ca^{2+}]_i$ in A431 cells. On the other hand, the absolute requirement for external Ca^{2+} could, of course, simply reflect a Ca^{2+}-dependent coupling of individual EGF receptors to the IP_3-generating process.

Inositol Lipid Breakdown

Much attention has recently been focused on the role of inositol 1,4,5-trisphosphate [IP_3 (1,4,5)] as a specific releaser of intracellular Ca^{2+} following receptor stimulation (Berridge and Irvine 1984). It is now becoming increasingly apparent that several additional inositol polyphosphates, with as yet unknown functions, are produced in stimulated cells. In particular, an IP_3 (1,3,4)-isomer has been detected which seems to be formed by dephosphorylation of inositol-1,3,4,5-tetrakisphosphate (IP_4, Batty et al. 1985). Furthermore, the evidence of

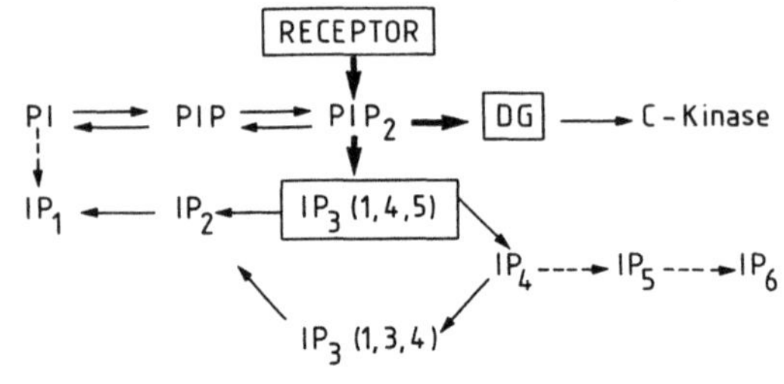

Fig. 4. Proposed metabolic pathways responsible for the formation of inositol phosphates. *PI(P)* phosphatidyl-inositol-(phosphate); IP_1 inositol-monophosphate; IP_2 inositol-bisphosphate. See text for further details

IP_5 and IP_6 has been described (Heslop et al. 1985). Figure 4 summarizes some of the known and proposed metabolic pathways responsible for the formation of inositol phosphates (cf. Heslop et al. 1985). Although it seems likely that IP_3 (1,3,4) and IP_3 may have second messenger functions, this idea remains to be tested.

We have separated ^3H-inositol phosphates from EGF-treated A431 cells using an HPLC anion-exchange system. Growing A431 cells were found to contain two IP_3 isomers at roughly equal concentrations, while three prominent peaks with decreasing ionophoretic mobilities were tentative identified as IP_4, IP_5 and IP_6 respectively (B.C. Tilly and P. van Paridon unpublished data). Addition of EGF causes a small but significant increase in the level of IP_3 (1,4,5), while there is a ~three-fold increase in IP_3 (1,3,4) concentration within 1 min (Table 1). These results, although preliminary, strengthen the view that the IP_3 (1,3,4)-isomer may have a second messenger role, possibly in mediating Ca^{2+} entry (see previous section). Time-course studies and micro-injection experiments should lead to the verification or disproval of this hypothesis.

Table 1. ^3H-inositol phosphate levels in A431 cells

	d.p.m. (x 10^{-2}) per sample	
	Control	EGF
IP_3(1,4,5)	11	18
IP_3(1,3,4)	3.5	11
IP_4	5.3	9
IP_5 + IP_6	117	83

Levels of inositol phosphates were measured in nearly confluent cultures pre-labeled to isotopic equilibrium with ^3H-inositol (2 µCi/ml). Treatment with EGF (100 ng/ml) lasted for 60 s. Cell extracts were processed for analysis of ^3H-inositol phosphates by an HPLC anion-exchange system (Tilly et al. in preparation)

Dissociation of Signal Pathways by Anti-Receptor Monoclonal Antibodies

The partial activation of post-receptor signaling pathways by phorbol esters has proved to be a fruitful approach in elucidating interrelationships among these pathways. An alternative tool for the dissociation of molecular events in the signaling cascade is provided by the availability of monoclonal antibodies to the EGF receptor. We have used three different anti-EGF receptor monoclonal IgG's, directed against distinct epitopes of the extracellular domain of the human EGF receptor (Defize et al. 1986), to test their ability to act as partial or full agonists of the EGF receptor. All three antibodies (named 2E9, 2D11, and 2G5, respectively) are able to immunoprecipitate a functional EGF receptor showing EGF-dependent tyrosine kinase activity. Monoclonal 2E9 is unique in that it recognizes a peptide determinant at or close to the EGF binding domain of the receptor. As a consequence, 2E9 competitively inhibits EGF receptor binding. In contrast, the other monoclonals ((2D11 and 2G5) are directed to bloodgroup A-specific carbohydrate structures on the EGF receptor and fail to affect EGF binding. We have tested these antibodies for their EGF-like properties in stimulating the receptor-mediated tyrosine phosphorylations, cytoplasmic alkalinization, Ca^{2+} mobilization and DNA synthesis. As summarized in Table 1, all three monoclonal IgG's can stimulate the tyrosine-specific autophosphorylation of the 170 kDa EGF receptor both in isolated A431 membranes and in intact cells (Defize et al. 1986). Interestingly, none of these antibodies is capable of triggering inositol phosphate formation and generation of ionic signals (Table 2), even after addition of a second cross-linking anti-IgG. Finally, the anti-receptor antibodies fail to stimulate DNA synthesis in quiescent human fibroblasts. Stimulation of the receptor's intrinsic tyrosine kinase is apparently not sufficient, by itself, to elicit a mitogenic response (for further details see Defize et al. 1986). Another important conclusion from those results is that stimulation of the EGF receptor kinase does not necessarily activate the post-receptor pathway that leads to phospholipase C activation and to an increase in $[Ca^{2+}]_i$ and in pH_i. These findings further support the view that the two major ionic signals are indispensable for the stimulation of DNA synthesis and cell proliferation.

Possible Physiological Role of the Ionic Signals

Cytoplasmic Alkalinization

An early rise in pH_i appears to be a fairly common response of metabolically dormant cells to appropriate surface stimuli. Several lines of evidence support the hypothesis that a sustained increase in pH_i, mediated by the Na^+/H^+ exchanger, may be required for mitogenesis. Perhaps the clearest demonstration of a role for cytoplasmic alkalinization in the initiation of a mitogenic response comes from studies on fertilized sea urchin eggs, where it has been shown that pH_i must rise from 6.8 to above 7.0 to permit DNA synthesis to begin (Whitaker and Steinhardt 1982). One of the critical pH_i-dependent steps in the fertilized eggs appears to be the stimulation of protein synthesis. From studies on somatic cells it is known that pH_i may participate in the regulation of such diverse processes as glycolosis cell-to-cell communication and cytoskeletal reorganization (reviewed by Busa and Nuccitelli 1984).

Figure 5 shows the effects of artificially shifting pH_i on the ability of quiscent human fibroblasts to reinitiate DNA synthesis in the presence of serum. It is seen that below a certain threshold value (near 7.0) pH_i becomes limiting for DNA synthesis, but that thymidine incorporation dramatically increases between pH_i 7.0-7.2. It is also observed that an artificially induced rise in pH_i is not sufficient, in itself, to stimulate DNA synthesis in serum-deprived cells (Fig. 5). These results agree, at least qualitatively, with earlier observations in thrombin-stimulated hamster fibroblasts: using mutant cells that lack a functional Na^+/H^+ exchanger, Pouysségur et al. (1984, 1985) elegantly showed that the initiation of DNA synthesis in mitogen-stimulated cells

Table 2. Comparison of the biological effects of EGF and anti-EGF receptor monoclonal antibodies on human A431 cells and fibroblasts

	EGF	2E9	2D11
Precipitation of EGF receptor		+	+
EGF binding competition	+	+	-
Stimulation of tyrosine kinase	+	+	+
Morphological changes	+	-	+
Inositol phosphate formation	+	-	-
Rise in $[Ca^{2+}]_i$	+	-	-
Rise in pH_i	+	-	-
Stimulation of DNA synthesis	+	-	-

Stimulation of DNA synthesis was tested on quiscent fibroblasts; all other effects on A431 cells. For further details see Defize et al. (1986).

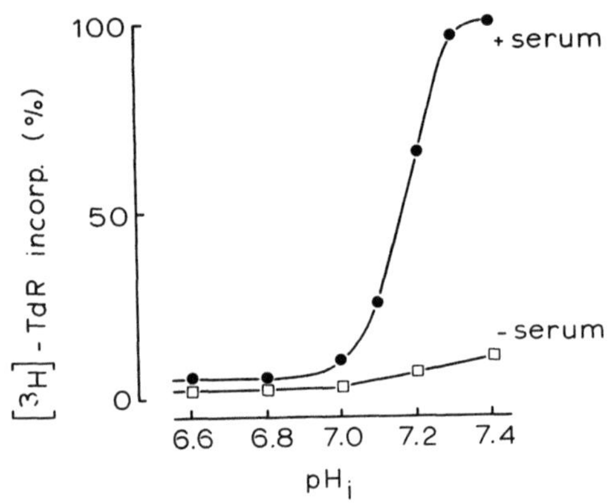

Fig. 5. Dependence of DNA synthesis on pH_i in human fibroblasts. Growth-arrested fibroblasts were incubated in HEPES-buffered media in the presence (●) or absence (□) of serum (10% v/v) and their pH_i was adjusted to the indicated values (by changing extracellular pH) for a period of 20 h. Thereafter, $[^3H]$-thymidine (1 µCi/ml) was added at physiological external pH (7.4) and acid-insoluble counts were determined at 24 h

is extremely sensitive to pH_i. At pH_i <7.2 DNA synthesis is completely blocked, but the mutant cells resume growth as soon as their pH_i is artificially elevated, either by raising external pH or incubating them in bicarbonate-buffered media. Addition of bicarbonate apparently raises pH_i into the "permissive range" due to the operation of a Na^+-dependent HCO_3^-/Cl^- exchanger (l'Allemain et al. 1985).

Increases in pH_i during the cell cycle have been observed in a few differrent cell types. In the slime mold *Physarum*, pH_i rises rapidly just before mitosis, whereas in *Tetrahymena* there are two alkaline pH_i transients per cell cycle (Busa and Nuccitelli 1984). In the cellular slime mold *Dictyostelium discoideum* there is a cycle of pH_i that closely corresponds to the DNA replication cycle, with a peak during S phase and mitosis (Aerts et al. 1985). Interestingly, when protein and DNA synthesis are blocked, pH_i continues to oscillate with a period similar to the uninhibited division cycle. Conversely, an artificial increase in pH_i by only 0.1 unit causes a severalfold increase in the rates of protein and DNA synthesis. It was concluded that this autonomous "pH_i oscillator" may have an important role in the timing and regulation of the cell cycle (Aerts et al. 1985). It was further hypothesized that the observed pH_i cycles are due to periodic alterations in the activity of a plasma membrane H+ pump, most likely an H+-ATPase.

Obviously, it would be of great interest to determine whether pH_i regulated by Na^+/H^+ exchange exhibits similar oscillations during the cell cycle of higher eukaryotes.

In summary, the available evidence provides strong support for the view of that pH_i plays an important role in the regulation of cellular metabolism in general and in the regulation of DNA synthesis and cell proliferation in particular.

Rise in $[Ca^{2+}]_i$

Another key question is what the physiological role of the $[Ca^{2+}]_i$ transient is in the action of growth factors. Numerous cellular functions are regulated by $[Ca^{2+}]_i$ and many of them through the formation of Ca^{2+}-calmodulin complexes. A discussion on this topic is beyond the scope of this brief review. Of particular relevance to mitogen action is the finding that artificially raising $[Ca^{2+}]_i$ by Ca^{2+} ionophore A23187 can mimic the effects of EGF and PDGF on the rapid induction of the c-*fos* and c-*myc* proto-oncogenes (Bravo et al. 1985, Tsuda et al. 1985). We recently found that Ca^{2+}-mobilizing neurohormones like bradykinin and histamine similarly induce c-*fos* and c-*myc* expression as well as cell division in their target cells (W. Kruijer and W.H. Moolenaar unpublished data). These results raise the intriguing possibility that Ca^{2+} could function as a second messenger that mediates, either directly or indirectly, the early transcriptional effects of growth factors and some hormones and they provide a framework for future studies on the role of intracellular Ca^{2+} in growth control.

Concluding Remarks

In recent years much has been learned about the various molecular events that are induced by growth factor-receptor interaction. These include tyrosine-specific protein phosphorylations, inositol lipid breakdown and the two major ionic events discussed above, namely, transient elevation of $[Ca^{2+}]_i$ and activation of Na^+/H^+ exchange resulting in a sustained rise in pH_i. There is good evidence that both ionic events are the direct consequence of the receptor-mediated hydrolysis of inositol phospholipids, while tyrosine kinase activity seems to initiate a distinct pathway, as summarized in the scheme of Fig. 6. Although uncertainty still exists about the biological significance of each of these "early" steps in the eventual initiaition of DNA synthesis, occurring many hours later, it seems likely that the various events evoked by inositol lipid breakdown act in concert with tyrosine-specific protein phosphorylations to stimulate cell proliferation. Selective pharmacological inhibitors without nonspecific side effects could greatly assist in elucidating the relevance of each of the signal pathways, but such agents are lacking at present. It is obvious that mutant cells defective in Na^+/H^+ exchange activity (Pouyssêsgur et al. 1984, 1985) or other functions, as well as anti-receptor monoclonal antibodies that act as partial agonst (Defize et al. 1986) are of enormous help in analyzing the role of the various signal pathways in growth control.

Another challenge for future studies is to examine whether the ionic signals generated by growth factor receptors have their correlates in the action of certain transforming oncogene products, some of which are known to be tyrosine-specific protein kinases. It is reasonable

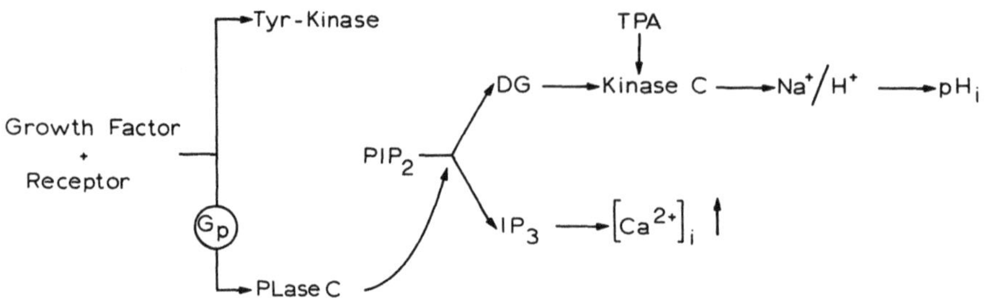

Fig. 6. Proposed seqeuence of events leading from growth factor binding to ionic signals. G_p is the putative GTP-binding protein; *Tyr-Kinase* tyrosine-specific protein kinase; *PLase C* phospholipase C; *PIP2* phosphatidylinositol 4,5-bisphosphate; *DG* 1,2-diacylglycerol; *IP3* inositol 1,4,5-trisphosphate; *TPA* 12-O-tetradecanoyl phorbol-13-acetate. See text for further details

to speculate that malignant growth, induced by inappropriate expression of cellular proto-oncogenes, is partially the result of the continual production of second messenger like IP_3 and diacylglycerol. This might lead to altered cellular Ca^{2+} pools and to constitutive activation of kinase C and, hence, of the Na^+/H^+ exchanger.

Acknowledgments. Research related to this review was supported by the Netherlands Cancer Foundation (koningin Wilhelmina Fonds) and the Organization for the Advancement of Pure Research (ZWO).

References

Aerts RJ, Durston AJ, Moolenaar WH (1985) Cytoplasmic pH and the regulation of the Dictyostelium cell cycle. Cell 43:653-657

Aronson PS, Nee J, Suhm MA (1982) Modifier role of internal H^+ in activating the Na-H exchanger in renal microvillus membrane vesicles. Nature (London) 299:161-163

Batty IR, Nahorski SR, Irvine RF (1985) Rapid formation of IP_4 following muscarinic stimulation of rat cerebral cortical slices. Biochem J 232:211-215

Berridge MJ, Irvine RF (1984) Inositol trisphosphate, a novel second messenger in cellular signal transduction. Nature (London) 312:315-321

Berridge MJ, Heslop JP, Irvine RF, Brown KD (1984) Inositol trisphosphate formation and calcium mobilization in Swiss 3T3 cells in response to PDGF. Biochem J 222: 195-201

Besterman JM, May WS, LeVine H, Cragoe EJ, Cuatrecasas P (1985) Amiloride inhibits phorbol ester-stimulated Na^+/H^+ exchange and protein kinase C. J Biol Chem 260: 1155-1159

Bishop JM (1985) Viral oncogenes. Cell 42:23-38

Bravo R, Burckhardt J, Curran T, Müller R (1985) Stimulation and inhibition of growth by EGF in different A431 cell clones is accompanied by the rapid induction of c-*fos* and c-*myc* proto-oncogenes. EMBO J 4:1193-1198

Busa WB, Nuccitelli R (1984) Metabolic regulation via intracellular pH. Am J Physiol 246:R409-R438

Carpenter G (1984) Properties of the receptor for epidermal growth factor. Cell 37: 357-358

Cockroft S, Gomperts BD (1985) Role of guanine nucleotide binding protein in the activation of polyphosphoinositide phosphodiesterase. Nature (London) 314:534-536

Davis RJ, Czech MP (1985) Amiloride directly inhibits growth factor receptor tyrosine kinase activity. J Biol Chem 260:2543-2551

Defize LHK, Moolenaar WH, Saag PT van der, Laat SW de (1986) Dissociation of cellular response to EGF using antireceptor monoclonal antibodies. EMBO J 5:1187-1192

Grinstein S, Cohen S, Goetz JD, Rothstein A, Gelfand EW (1985) Characterization of the activation of Na^+/H^+ exchange in lymphocytes by phorbol esters: Change in cytoplasmic pH dependence of the antiport. Proc Natl Acad Sci USA 82:1429-1433

Heldin C-H, Westermark B (1984) Growth factors: Mechanism of action and relation to oncogenes. Cell 37:9-20

Hesketh TR, Moore JP, Morris JDH, Taylor MV, Rogers J, Smith GA, Metcalfe JC (1985) A common sequence of calcium and pH signals in the mitogenic stimulation of eukaryotic cells. Nature (London) 313:481-484

Heslop JP, Irvine RF, Tashjian AT, Berridge MJ (1985) Inositol tetrakis-and pentakisphosphates in GH_4 cells. J Exp Biol 119:396-401

Hunter T, Cooper JA (1985) Protein-tyrosine kinase. Annu Rev Biochem 54:897-93o

L'Allemain G, Franchi A, Cragoe EJ, Pouysségur J (1984) Blockade of the Na^+/H^+ antiport abolishes growth factor-induced DNA synthesis in fibroblasts. J Biol Chem 259:4313-4319

L'Allemain G, Paris S, Pouysségur J (1985) Role of Na^+-dependent Cl^-/HCO_3^- exchange in regulation of intracellular pH in fibroblasts. J Biol Chem 260:4877-4883

Molski TFP, Naccache PH, Volpi M, Wolpert LM, Sha'afi RI (1980) Specific modulation of the intracellular pH of rabbit neutrophils by chemotactic factors. Biochem Biophys Res Commun 94:508-514

Moolenaar WH (1986a) Effects of growth factors on intracellular pH regulation. Annu Rev Physiol 48:363-376

Moolenaar WH (1986b) Regulation of cytoplasmic pH by Na^+/H^+ exchange. Trends Biochem Sci 11:141-143

Moolenaar WH, Laat SW de, Saag PT van der (1979) Serum triggers a sequence of rapid ionic conductance changes in quiescent neuroblastoma cells. Nature (London) 279: 721-723

Moolenaar WH, Mummery CL, Saag PT van der, Laat SW de (1981a) Rapid ionic events and the initiation of growth in serum-stimulated neuroblastoma cells. Cell 23: 789-798

Moolenaar WH, Boonstra J, Saag PT van der, Laat SW de (1981b) Sodium/proton exchange in mouse neuroblastoma cells. J Biol Chem 256:12883-12887

Moolenaar WH, Yarden Y, Laat SW de, Schlesinger J (1982) Epidermal growth factor induces electrically silent Na^+ influx in human fibroblasts. J Biol Chem 257: 8502-8506

Moolenaar WH, Tsien RY, Saag PT van der, Laat SW de (1983) Na^+/H^+ exchange and cytoplasmic pH in the action of growth factors in human fibroblasts. Nature (London) 304:645-648

Moolenaar WH, Tertoolen LGJ, Laat SW de (1984) Growth factors immediately raise cytoplasmic free Ca^{2+} in human fibroblasts. J Biol Chem 259:8066-8069

Moolenaar WH, Aerts RJ, Tertoolen LGJ, Laat SW de (1986) The epidermal growth factor-induced calcium signal in A431 cells. J Biol Chem 261:279-284

Nishizuka Y (1984) The role of protein kinase C in cell surface signal transduction and tumour promotion. Nature (London) 308:693-698

Ohsako S, Deguchi T (1981) Stimulation by phosphatidic acid of calcium influx and cyclic GMP synthesis in neuroblastoma cells. J Biol Chem 256:10945-10948

Paris S, Poysségur J (1984) Growth factors activate the Na^+/H^+ antiporter in quiescent fibroblasts by increasing its affinity for intracellular H^+. J Biol Chem 259:10989-10994

Paris S, Pouysségur J (1986) Pertussis toxin inhibits thrombin-induced activation of phosphoinositide hydrolysis and Na^+/H^+ exchange in hamster fibroblasts. EMBO J 5:55-60

Pouysségur J, Sardet LC, Franchi A, l'Allemain G, Paris S (1984) A specific mutation abolishing Na^+/H^+ antiport activity in hamster fibroblasts precludes growth at neutral and acidic pH. Proc Natl Acad Sci USA 81:4833-4837

Pouysségur J, Franchi A, l'Allemain G, Paris S (1985) Cytoplasmic pH, a key determinant of growth factor-induced DNA synthesis in quiescent fibroblasts. FEBS Lett 190:115-118

Putney JW, Weiss SJ, Walle CM van de, Haddas RA (1980) Is phosphatidic acid a calcium ionophore under neurohumoral control? Nature (London) 284:345-347

Roos A, Boron W (1981) Intracellular pH. Physiol Rev 61:296-434
Rothenberg P, Reuss L, Glaser L (1982) Serum and EGF transiently depolarize quiescent BSC-1 epithelial cells. Proc Natl Acad Sci USA 79:7783-7787
Rozengurt E (1981) Stimulation of Na^+ influx, Na^+, K^+ pump activity and DNA synthesis in quiescent cultured cells. Adv Enzyme Regul 19:61-85
Sagi-Eisenberg R, Lieman H, Pecht I (1985) Protein kinase C regulation of the receptor-coupled calcium signal in histamine-secreting rat basophilic leukaemia cells. Nature (London) 313:59-60
Sahai A, Smith KB, Pannerselvam M, Salomon DS (1982) Activation of calcium and phospholipid-dependent protein kinase by EGF in A431 cells. Biochem Biophys Res Commun 109:11206-1214
Sawyer ST, Cohen S (1981) Enhancement of calcium uptake and phosphatidylinositol turnover by EGF in A431 cells. Biochemistry 20:6280-6286
Serhan CN, Fridovich J, Goetzl EJ, Dunham P, Weissmann G (1982) Leukotriene B_4 and phosphatidic acid are calcium ionophores. J Biol Chem 257:4746-4752
Tsuda T, Kaibuchi K, West B, Takai Y (1985) Involvement of Ca^{2+} in platelet-derived growth factor-induced expression of c-myc oncogene in Swiss 3T3 fibroblasts. FEBS Lett 187:43-46
Vara F, Rozengurt E (1985) Stimulation of Na^+/H^+ antiport activity by EGF and insulin occurs without activation of protein kinase C. Biochem Biophys Res Commun 130:646-653
Whitaker MJ, Steinhardt RA (1982) Ionic regulation of egg activation. Q Rev Biophys 15:593-666
Whiteley B, Cassel D, Zhunang Y, Glaser L (1984) The tumor promoter 12-myristate 13-acetate inhibits mitogen-stimulated Na^+/H^+ exchange in human epidermoid carcinoma A431 cells. J Cell Biol 99:1162-1166
Zhuang Y, Cragoe EJ, Shaikewitz T, Glaser L, Cassel D (1984) Characterization of potent Na^+/H^+ exchange inhibitors of the amiloride series in A431 cells. Biochemistry 23:4481-4488

Structural and Functional Analysis of Cell Surface Receptors with Tyrosine Kinase Activity

H. Riedel[1], Y. Yarden[1], L. Coussens[1], J. Lee[1], T. J. Dull[1], A. Gray[1], J. Schlessinger[2], and A. Ullrich[1]

1 Introduction

Communication between specialized cells is an essential component of multicellular biological systems. Secreted polypeptide hormones and specific receptor molecules present on the surfaces of target cells represent key components in this communication network. The interaction of hormones such as insulin, insulin-like growth factor I (IGF-I), epidermal growth factor (EGF), platelet-derived growth factor (PDGF) and colony-stimulating factor I (CSF-I) with the extracellular portions of their cell surface receptors triggers a distinct cellular response, which may differ depending on the type and developmental stage of the target cell (Kahn 1985; Carpenter 1981). We have undertaken a molecular biological approach as a first step towards unravelling the cytoplasmic and nuclear events that are initiated by extracellular complex formation between ligands and their respective cell surface receptors. As a first step in this undertaking we characterized the primary sequences of several receptors, hoping to obtain clues regarding the overall domain structure of these cell surface molecules and their possible involvement in ligand-induced signal generation. Furthermore we are using these cloned receptor cDNA's for the generation and testing of chimeric molecules in an attempt to define receptor domains and their functions.

2 Structural Classification of Receptor Tyrosine Kinases

Comparison of cDNA-derived sequences of EGF (Ullrich et al., 1984), insulin (Ullrich et al., 1985; Ebina et al. 1985), IGF-I (Ullrich et al. 1986), PDGF (Yarden et al. 1986), and CSF-I (Coussens et al. 1986) receptors, and the EGF receptor related putative receptor HER2 (Coussens et al. 1985) reveals common characteristics and distinct structural features that subdivide this receptor family into three groups (Fig. 1). Common features include a single stretch of predominantly hydrophobic amino acids representing a transmembrane anchor sequence, an amino-terminal extracellular domain that forms the ligand binding pocket, and an intracellular domain that includes sequences specifying the tyrosine-specific protein kinase activity.

Distinctive structural characteristics define three receptor types with tyrosine kinase activity (Table 1), including EGF receptor and HER2 (type 1), insulin receptor (A and B variants, see legend to Fig. 1) and IGF-I receptor (type 2), and PDGF and CSF-1 receptors (type 3).

[1]Department of Developmental Biology, Genentech, Inc., 460 Point San Bruno Boulevard, South San Francisco, California 94080, USA

[2]Biotechnology Research Center, Meloy Laboratories, 4 Research Court, Rockville, Maryland 20850, USA

37. Colloquium Mosbach 1986
Cell Cycle and Oncogenes

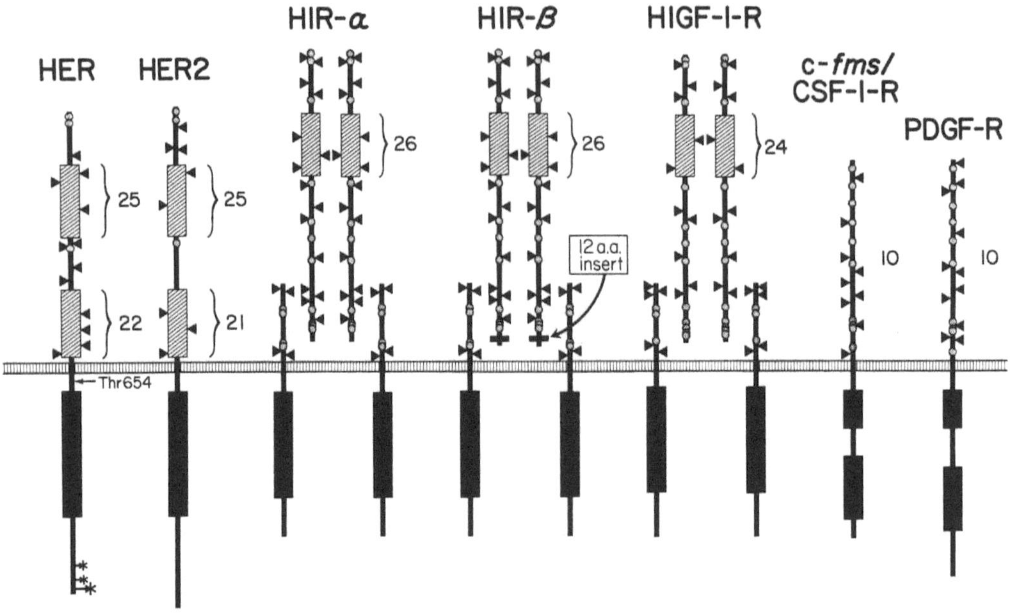

Fig. 1. Schematic presentation of receptor structures. Human EGF receptor (HER), EGF receptor related receptor (HER2), insulin receptor (HIR α and β variants, which differ by the presence of 12 additional amino acids in β), IGF-I receptor (HIGF-I-R), CSF-I receptor (CSF-I-R) and mouse PDGF receptor (PDGF-R) are shown inserted in the plasma membrane bilayer with their amino-terminal domains outside and their carboxy-terminal domains inside the cell. Cysteine-rich clusters (*shaded boxes;* with numbers of cysteine residues indicated), potential N-linked glycosylation sites (*black triangles*), cysteine residues (*dotted circles*) and tyrosine kinase domains (*black boxes*) are indicated. Amino acids involved in regulation of EGF receptor (HER1) action (Thr 654) and in vitro autophosphorylation sites (*asterisks*) are indicated.

Table 1. Family of receptor tyrosine kinases

Type I	Type II	Type III
EGF receptor	Insulin receptor (α and β subtypes)	c-*fms*/CSF-1 receptor
HER2/*neu*	IGF-1 receptor	PDGF receptor
	Characteristics	
Monomer	Heterotetramer	Monomer
2 Cys-rich extracellular sequence repeat domains	1 Cys-rich extracellular sequence repeat domain per alpha subunit	10 extracellular cyxteines
Basic juxtamembrane domain		Split tyrosine kinase domain

While the type 1 and type 3 receptors are synthesized and function as single polypeptide chains, mature type 2 receptors are generated by cleavage of a proreceptor precursor of α and β subunits, which are then combined by disulfide bridges to form an $\alpha_2\beta_2$ heterotetrameric complex. Extracellular ligand binding domains of type 1 and type 2 receptors share the structural motif of conserved cysteine-rich sequence repeat clusters. Two of these subdomains are found in EGF receptor and HER2, while only one is present in the insulin and IGF-I receptor α subunits which, however, results in two per heterotetrameric complex (Ullrich et al. 1984,1985,1986). In contrast, ten cysteine residues are found in the extracellular domain of type 3 receptors for PDGF and CSF-1, which are conserved with respect to their position. The first six cysteine residues are separated by similar distances (45-50 amino acids) and are flanked by related sequences, suggesting the possibility of a sequence multiplication event during evolution of these receptor genes (Coussens et al. 1986; Yarden et al. 1986).

These differences in extracellular domain structures indicate different architectural principles used for the formation of ligand binding pockets and reflect possible differences in molecular characteristics of respective polypeptide ligands. Furthermore, it suggests the possibility that type 1 and 2 receptors originated in an evolutionary event independent from the one that generated the type 3 subgroup. This speculation is supported by an additional structural feature characteristic of type 3 receptors. Both PDGF and CSF-I receptor tyrosine kinase domains are split by an insertion sequence of approximately 100 and 70 amino acids, respectively, while type 1 and 2 receptors show only minor (8-14 amino acids) divergent sequences in this position.

3 Relationship Between Oncogene Products and Receptor Tyrosine Kinases

Until recently, the role of normal cellular homologues of transforming genes was poorly understood. Some of these genes have now been shown to encode growth factor receptors, which when constitutively expressed, structurally altered or expressed in an inappropriate cell type may result in subversion of growth control mechanisms and thus oncogenesis. Four members of the tyrosine kinase oncogene family, v-*erbB*, v-*fms*, v-*ros* and *neu*, have been found to be highly homologous to the known or putative growth factor receptors (Fig. 2). The transforming gene of the avian erythroblastosis virus v-*erbB* is believed to be an altered derivative of the avian EGF receptor (Downward et al. 1984a; Ullrich et al. 1984), the receptor for macrophage colony-stimulating factor CSF-1 may be identical or related to the feline oncogene product v-*fms* (Sherr et al. 1985), the transforming gene of the avian sarcoma virus UR-2 v-*ros* (Neckameyer and Wang, 1984) shows sequence similarity with insulin and IGF-I receptors (Ullrich et al. 1985, 1986), and the chemically activated *neu* oncogene product has been shown to be structurally similar to the EGF receptor and v-*erbB* (Coussens et al. 1985; Bargmann et al. 1986; Yamamoto et al. 1986) and most likely represents another cell surface receptor for an unknown ligand with as yet unknown biological activity.

While v-*erbB* and v-*ros* appear to be amino- and carboxy-terminally truncated derivatives of their respective receptor homologues, v-*fms* still retains an intact extracellular ligand binding domain, but lacks 40 carboxy-terminal amino acids. In all cases the carboxyl termini of cellular cognates contain tyrosine residues, which may have an important role in signal transduction. A tyrosine residue is found at position 969 of c-*fms* in a location similar to that for the EGF receptor tyro-

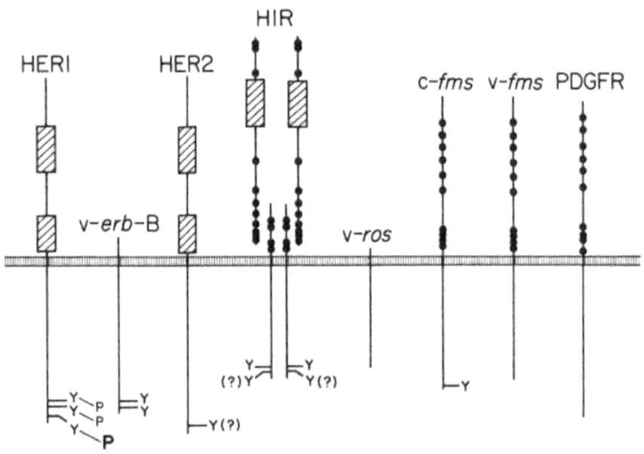

Fig. 2. Schematic comparison of structural features of cell surface receptors and oncogene products. Regions of high-Cys concentration are shown as *hatched boxes* and single Cys residues as *filled circles*. The position of carboxy-terminal tyrosine residues is shown. *HIR* human insulin receptor; *HERI* human epidermal growth factor receptor; *HER2*, *EGF* receptor-related receptor

sine 1173, which is a major autophosphorylation substrate for the ligand-induced tyrosine autophosphorylation activity (Downward et al. 1984b). The difference between v- *fms* and v-*ros*/v-*erbB* may reflect alternative mechanisms used by the respective receptor molecules to control transmembrane signal generation. Double truncation of cellular *ros* and *erbB* proto-oncogenes may be necessary to generate a transforming gene, while a single carboxyl-terminal truncation of the c-*fms* gene may yield the same result. Alternatively, the transforming potential of v- *fms* may only be activated by an autocrine mechanism in cells secreting the CSF-I ligand.

4 Reconstitution of Ligand Responsiveness in a Chimeric *erbB/EGF* Receptor

Comparison of the complete deduced amino acid sequence of the human EGF receptor (Ullrich et al. 1984) with the avian erythroblastosis virus *erbB* oncogene product (Yamamoto et al. 1983) revealed close similarity, and suggested that v-*erbB* was derived from the avian EGF receptor gene by gene truncation as a consequence of the recombination event that created the AEV-H virus strain. Both molecules are transmembrane tyrosine kinases, and when aligned with the human EGF receptor sequence, v-*erbB* lacks the putative signal peptide for membrane translocation (24 amino acids), most of the extracellular ligand binding domain (amino acids 1-555), and 32 carboxy-terminal amino acids. Sequence homology is most extensive in the tyrosine kinase domain and fades towards the carboxy terminus beginning at residue 972 of the human EGF receptor (Ullrich et al. 1984).

To investigate the functional significance of structural features unique to the avian v-*erbB* oncogene, we constructed a carboxyl-terminal 32 amino acid truncation mutant of the human EGF receptor (HER6) as well as a chimeric protein comprised of the v- *erbB* intracellular domain fused to extracellular and transmembrane domains of the human EGF receptor (HER- *erbB*; Fig. 3). The sequence fusion was carried out without sequence alterations. The carboxy-terminal 32 amino acid deletion of HER6 was generated by oligonucleotide primer-directed introduction of a termination codon between residues 1154 and 1155 of the human EGF receptor sequence (Ullrich et al. 1984). DNA sequences coding for HER6- *erbB* were introduced into an expression vector containing the transcription start signals of the early region of simian virus

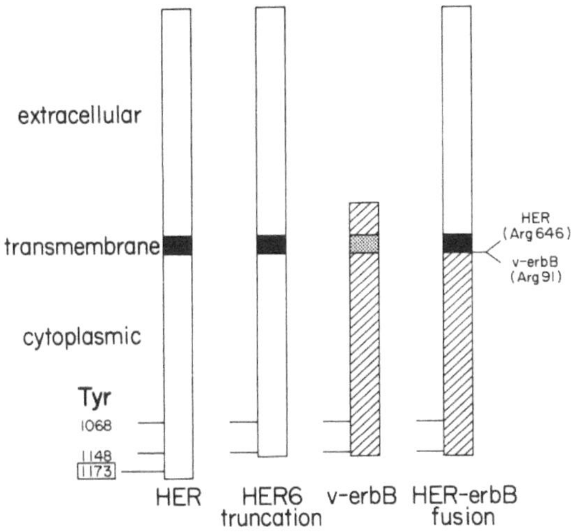

Fig. 3. Schematic comparison of truncation (*HER6*) and chimeric (*HER-erbB*) mutants. Molecules have been aligned at their transmembrane domain. AEV-*erbB* sequences are *shaded*. The major tyrosine in vitro autophosphorylation sites are indicated by *bars*

(SV) 40 or the Harvey sarcoma virus long terminal repeat, respectively. Signals in the 3'-untranslated region for the termination of transcription and the messenger RNA polyadenylation were provided by sequences of the gene coding for the hepatitis virus B surface antigen. The vector contained a second expression module containing a dihydrofolate reductase (DHFR) gene to permit gene amplification under methotrexate selection (Alt et al. 1978) and the HER-*erbB* plasmid in addition to a neomycin resistance gene (Colbere-Garapin et al. 1981; Southern and Berg 1982) for the initial selection of transfectants. Sequences of the *E. coli* plasmid pBR322 allowed plasmid DNA replication in *E. coli*.

Normal RatI fibroblasts were transfected by the calcium phosphate precipitation method (Graham and Van der Eb 1973) and after selection for neomycin (Neo) resistance, the introduced plasmid sequences were amplified under methotrexate (MTX) selection. Three cell lines, HER6/MTX, HER-*erbB*/Neo and HER-*erbB*/MTX, were used for subsequent experiments, and were compared with A431 epidermoid carcinoma cells which overexpress the normal EGF receptor (1-2 x 10^6 molecules/cell).

Expression of mutant constructs was monitored by immunoprecipitation of cell extracts after metabolic labeling using a human EGF receptor-specific monoclonal antibody (RI; Waterfield et al. 1982). The proteins migrated slightly faster than the human EGF receptor glycoprotein of A431 cells in reducing SDS polyacrylamide gels, suggesting comparable levels of carbohydrate addition to the primary translation products. When compared with the high level of EGF receptor expressed in these carcinoma cells, amplified RatI cells expressed only three-fold less HER-*erbB* and about six-fold less HER6.

The ability of transfected cells to bind increased levels of ^{125}I-EGF was used to determine whether mutant EGF receptors were properly transported to the cell surface of RatI recipient cells. All three transfected cell lines displayed dramatically increased ^{125}I-EGF surface binding relative to control RatI cells. ^{125}I-EGF binding could be competet with cold EGF and appears to correlate with protein levels detected by immunoprecipitation. Thus the constructed proteins contained fully functional EGF binding domains and were faithfully transported to the cell surface.

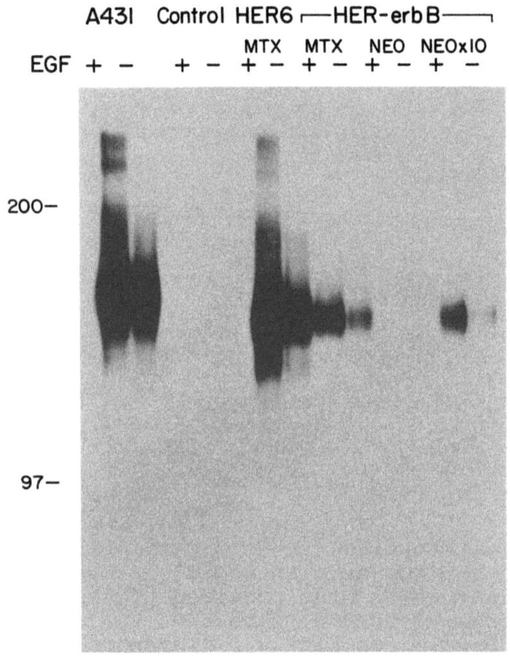

Fig. 4. EGF-stimulated autophosphorylation of mutant receptors. Proteins were immunoprecipitated from detergent lysates of stably expressing cells in the presence (+) or absence (-) of EGF. Reactions were analyzed by SDS polyacrylamide gel electrophoresis after 30 sec. at 4°C with ^{32}P-γ-ATP. The HER-*erbB* expression product is shown at levels comparable to the other constructs; in addition a tenfold higher concentration (*x 10*) is also presented. Normal RatI fibroblasts were used as a control. The employed monoclonal antibody (RI) does not recognize the endogenous RatI cell EGF receptor. Size markers are indicated in kilodaltons

To test whether the mutant receptor proteins HER6 and HER- *erbB* possessed in vitro autophosphorylation activity, immunoprecipitates were incubated with ^{32}P-γ-ATP and analyzed by SDS polyacrylamide gel electrophoresis and autoradiography. Like the A431 cell EGF receptor, both the HER6 truncation mutant and the HER-*erbB* chimeric protein incorporated ^{32}P by autophosphorylation in immunoprecipitates (Fig. 4). When compared with the A431 cell receptor, the specific autophosphorylation activity of the HER6 deletion mutant was reduced. This may be due to the absence of tyrosine residue 1173, which was found to be the major EGF receptor in vitro autophosphorylation substrate site (Fig. 1) (Downward et al. 1984b) or to an alteration in kinase activity caused by removal of 32 carboxy-terminal amino acids. However, the autophosphorylation activity of HER-*erbB* was even lower than that of HER6, which is consistent with the previously described low autophosphorylation activity of the v- *erbB* oncogene product (Kris et al. 1985), and may be due to cytoplasmic sequence differences between human and avian EGF receptors (Ullrich et al. 1984; Yamamoto et al. 1983). Both the truncated EGF receptor HER6 and the human-avian EGF receptor-oncogene chimera autophosphorylation activities were stimulatable by EGF, as was the A431 receptor control (Fig. 4).

Functional characterization of HER6 and HER- *erbB* mutants were tested by monitoring the mitogenic response of stably transfected RatI fibroblasts to EGF exposure. As shown in Fig. 5, control RatI fibroblasts displayed EGF-dependent thymidine incorporation due to endogenous rat EGF receptors. RatI cells expressing the HER6 deletion mutant showed enhanced thymidine incorporation in response to increasing EGF concentrations relative to control RatI cells. In contrast, expression of the HER- *erbB* chimera inhibited the mitogenic response mediated by the endogenous rat EGF receptor. The inhibitory effect was even more pronounced in the methotrexate amplified RatI cell line (HER-*erbB*/MTX) that expressed about eight-fold more of the chimeric EGF receptor mutant.

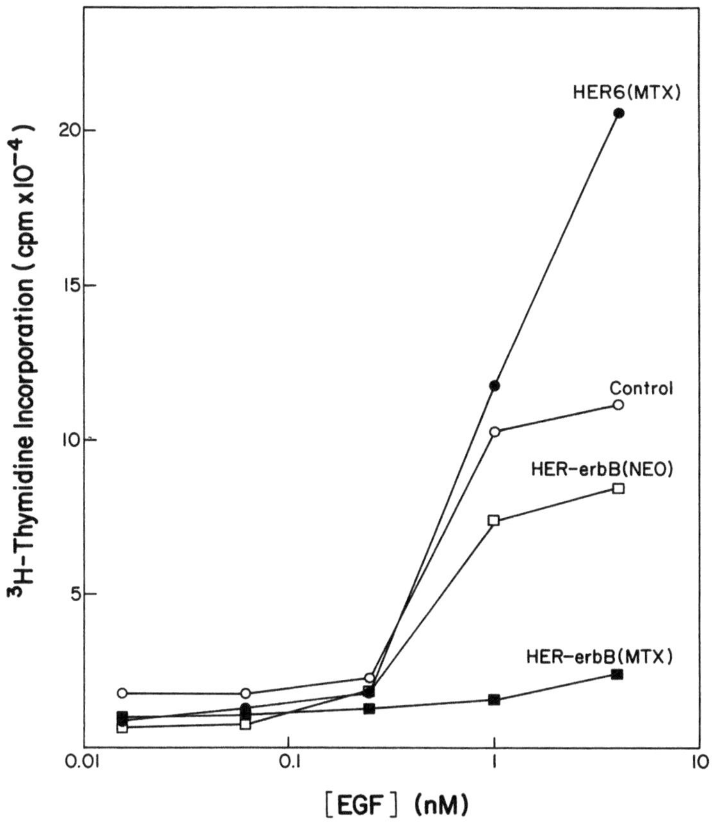

Fig. 5. Mitogenic response of transfected RatI fibroblasts to EGF. Stimulation of incorporation of ^{3}H-thymidine into DNA by EGF is shown for RatI cell lines expressing chimeric receptor and deletion mutants

To examine whether the *erbB* portion of the HER-*erbB* chimera lost its transforming potential and, if not, whether it is now under EGF control, cells were plated at high density and subsequently stimulated by EGF. Both growth inhibition and cell focus formation were observed for cells expressing the HER-*erbB* expression product (Fig. 6). This only occurred when an almost confluent monolayer of stably transfected RatI fibroblasts was exposed to EGF for several days in culture. Focus formation for HER- *erbB* was more pronounced at lower expression levels (HER- *erbB* /Neo) than at higher levels (HER/*erbB* /MTX), most likely due to cytotoxic consequences seen in the latter case. No cell foci were observed in the absence of EGF or under any conditions after expression of HER6 or for control RatI or A431 cells.

Anchorage-independent growth was tested as an independent measure of the transformation potential of the mutants. Only cells expressing high levels of HER -*erbB* formed colonies in soft agar. No colonies of control RatI fibroblasts or HER6 expressing cells were detected when plated in soft agar. It therefore appears that HER-*erbB* as well as v-*erbB* but not HER6 display a low cell transformation potential in the RatI cell system. EGF stimulation of soft agar colony formation could not be investigated because EGF induced soft agar growth even for control RatI fibroblasts.

The experiments demonstrate that avian v-*erbB* oncogenic sequences which are closely related to the human EGF receptor cytoplasmic domain

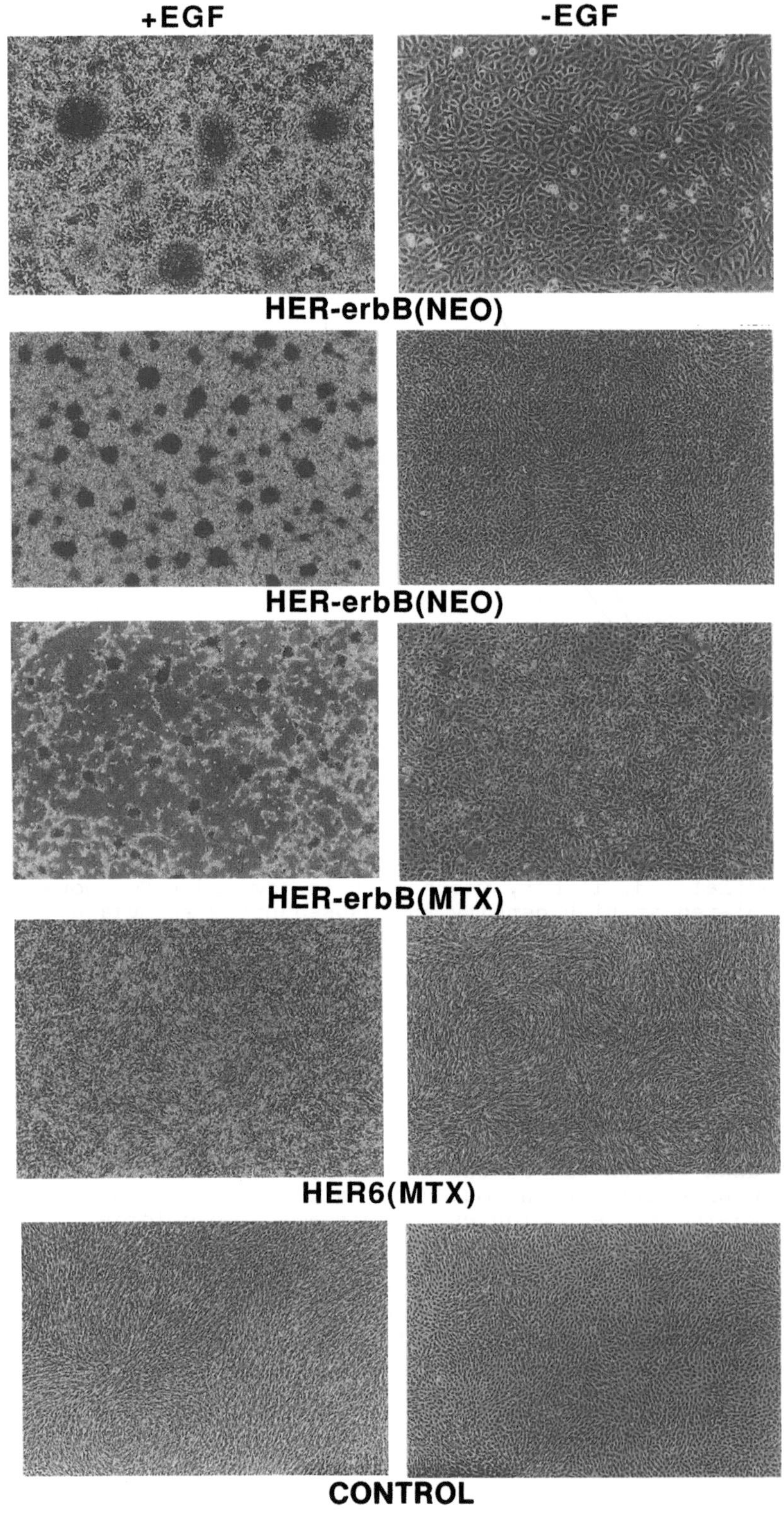
+EGF
-EGF
HER-erbB(NEO)
HER-erbB(NEO)
HER-erbB(MTX)
HER6(MTX)
CONTROL

can be recombined with a human EGF receptor ligand binding and transmembrane domain to yield a hybrid receptor in which the avian tyrosine kinase domain can be stimulated by EGF. Thus control of cytoplasmic activities by the extracellular receptor domain is reconstituted in this chimeric molecule. Comparison with a human EGF receptor deletion mutant HER6 lacking a stretch of 32 carboxy-terminal amino acids, including the major in vitro autophosphorylation substrate tyrosine residue (Tyr 1173), showed that the presence of these C-terminal sequences does not impair inducibility of the autophosphorylation activity. Reduced specific autophosphorylation activities, however, were observed for both the HER-*erbB* chimera and HER6 deletion mutant. This may reflect the loss of the major autophosphorylation target residue in both molecules, as well as a species-specific enzymatic feature characteristic of intact avian and human EGF receptors. Structural characterization of avian and human, oncogenic and normal EGF receptor sequences by Yamamoto et al. (1983), Nilsen et al. (1985) and Ullrich et al. (1984) revealed extensive homology between human and avian sequences within the tyrosine kinase core region (96%) and substantially increased sequence divergence towards the carboxyl terminus (Ullrich et al. 1984). The observed difference in specific autophosphorylation activity may therefore be defined by sequences outside this 327 amino acid region starting at the cytoplasmic border of the receptor, the transmembrane domain (Arginine 645 of the human EGF receptor). Influences by the heterologous extracellular and transmembrane domains or activity alterations by mostly conservative sequence differences (except EGF receptor Leucine 679 → Histidine and Glutamine 952 → Proline) cannot be excluded, although difficulties in detecting the weak v-*erbB* tyrosine kinase activity suggest this to be an avian EGF receptor characteristic.

The attempt to investigate whether the human deletion mutant and HER-*erbB* chimera confer a mitogenic response to RatI cells and whether the difference in enzymatic specific activity between human and avian cytoplasmic receptor sequences bears consequences for this key biological function resulted in the surprising finding that while HER6 expressing cells exhibited an enhanced DNA synthesis response, HER-*erbB* expression by RatI cells has an inhibitory effect on the mitogenic response mediated by the endogenous rat EGF receptor.

Our results further indicate that EGF binding to the receptor extracellular domain has a different cytoplasmic consequence than deletion of the binding domain, which in combination with the C-terminal deletion results in the transforming activity of v-*erbB*. The inhibitory effect on DNA synthesis, on the other hand, must be caused by structural determinants of HER-*erbB* which are different from the HER6 analog and which may be involved in intermolecular contacts necessary for the generation of a mitogenic signal. Such sequence differences (38%) are concentrated downstream from the end of the tyrosine kinase core (Fig. 7), suggesting that this subdomain of the EGF receptor is involved in generation of a mitogenic signal.

Human sequences present in HER6 are apparently compatible with necessary molecules in the rat fibroblast cell background, while respecti-

◀ Fig. 6. EGF-induced cell focus formation. Cells were plated at 20% confluency and grown in normal culture medium in the presence (+) or absence (-) of 100 ng/ml EGF for 1 week. Normal RatI fibroblasts were used as a control

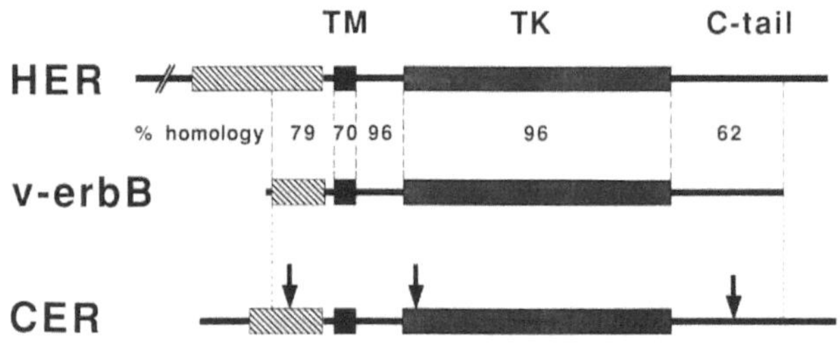

Fig. 7. Schematic comparison of human EGF receptor (*HER*), v-*erbB* and chicken EGF receptor (*CER*) sequences. Cysteine-rich clusters (*shaded boxes*), transmembrane domains (*TM*), and tyrosine kinase (*TK*) domains are emphasized. Extent of sequence homology between HER and *erbB* is shown in percent. *Arrows* indicate positions of single amino acid sequence differences between *erbB* and CER

ve avian structural determinants do not interact properly with rat cell factors and on the contrary act as competitive analogues, which results in inhibition of the endogenous mitogenic response of the cell. The finding that methotrexate-amplified RatI cells overexpressing the HER-*erbB* chimera show enhanced sensitivity to DNA synthesis inhibition demonstrates directly the causative connection between the HER-*erbB* molecule and physiological consequences observed.

Competition for cellular factors between the endogenous rat EGF receptor and HER-*erbB* or formation of nonfunctional dimers between HER-*erbB* and rat EGF receptor may explain the observed inhibitory effects. The latter possibility is based on the proposal of Yarden and Schlessinger (1986) that receptor dimers represent a functional unit. Structural differences within the same receptor region may also be responsible for the remaining transforming activity observed for HER-*erbB* expressing cells. When measured as focus formation in cell monolayers, this activity is EGF-inducible and is paralleled by the cytotoxic effect in overexpressing HER-*erb*/MTX cells, which does not allow formation of large foci. In addition, a residual, weak, EGF-independent colony formation activity in soft agar can be seen for HER-*erb* expressing RatI cells.

5 Insulin Stimulates the EGF Receptor Kinase in a Chimeric Receptor

To examine the potential of the receptor chimera approach for the characterization of structural domains involved in definition of receptor specific signalling functions, we designed a hybrid receptor molecule comprised of the extracellular portion of the insulin receptor joined to the transmembrane and intracellular domains of the EGF receptor (Fig. 8).

Cloned human EGF receptor (Ullrich et al. 1984) and insulin receptor cDNA's (Ullrich et al. 1985; Ebina et al. 1985) were used in combination with SV40 early promoter transcriptional elements to generate expression constructs coding for the complete human insulin receptor (HIR) and a chimeric receptor (IER), shown schematically in Fig. 8. The insulin receptor is normally synthesized as a receptor precursor that is proteolytically processed to generate the α and β subunits (Ullrich et al. 1985). IER was generated by fusing the coding region of the signal sequence and extracellular portion of the insulin receptor precursor [amino acids (-27) -914] to the coding sequence of the transmembrane and cytoplasmic domains of the EGF receptor (Ullrich et al. 1984) (amino acids 622-1186). Using this chimeric construct we asked whether proteolytic processing of the encoded hybrid receptor into insulin receptor α subunit and the recombinant βIE subunit could

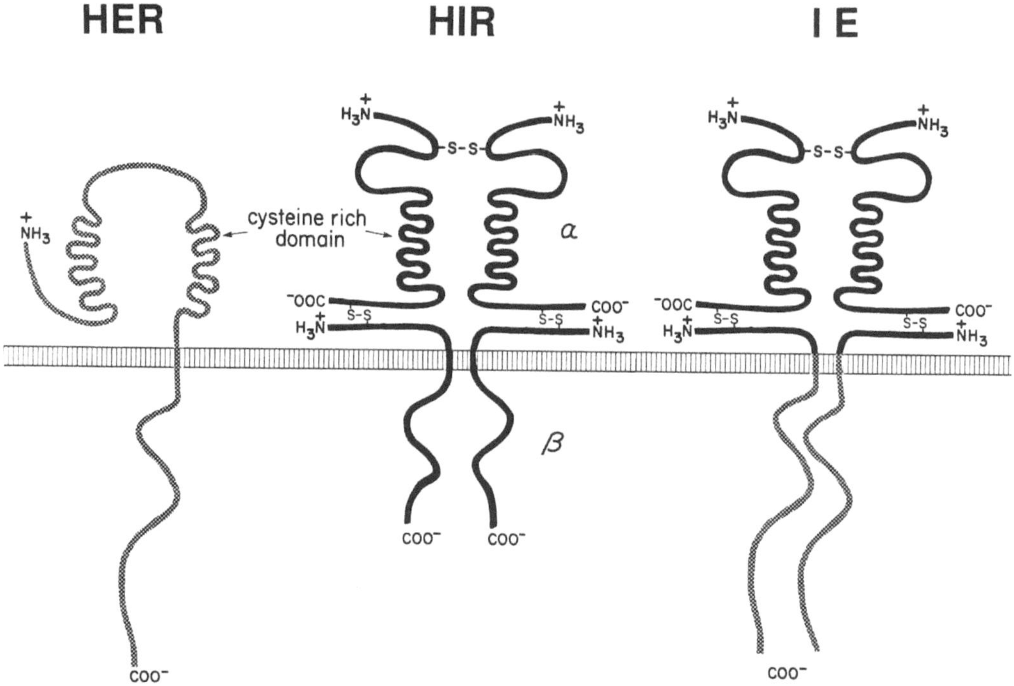

Fig. 8. Schematic display of human EGF receptor (*HER*), human insulin receptor (*HIR*), and insulin-EGF receptor chimera (*IE*) structures

occur, whether a functional complex would be formed and directed to the plasma membrane, and whether insulin binding could stimulate autophosphorylation of the EGF receptor cytoplasmic domain in the chimeric cell surface membrane glycoprotein.

Expression of the complete insulin receptor resulted in protein products of 95,000 M_r, 180,000 M_r, and a diffuse band around 135,000 M_r, as determined by ^{35}S-methionine metabolic labeling. These polypeptides represent the α-subunit, the β-subunit and the uncleaved precursor of insulin receptor (Ullrich et al. 1985; Massague et al. 1981; Kasuga et al. 1981; Reed et al. 1981; Deutsch et al. 1983). Proteolytic processing of the native insulin receptor precursor is inefficient under transient expression conditions in COS-7 cells, as evidenced by the amount of HIR precursor detected.

By analogy with the native receptors, COS-7 cells expressing the IER receptor chimera would be expected to synthesize a 135,000 M_r glycosylated α-chain, a comparably sized 120,000 M_r glycopolypeptide representing EGF receptor intracellular and transmembrane sequences fused to the insulin receptor β subunit extracellular portion (IβE), and a 190,000 M_r polypeptide, if cleavage of IER proreceptor sequences was inefficient. Each of these products was detected in cells expressing the IER chimera; the IβE fusion glycoprotein subunit migrated as a diffuse band just below the 135,000 M_r α-subunit polypeptide.

Binding experiments with ^{125}I-labeled insulin indicated that the products of the HIR and IER constructs were expressed in a functional form on the surface of COS cells. Insulin binding was the highest for

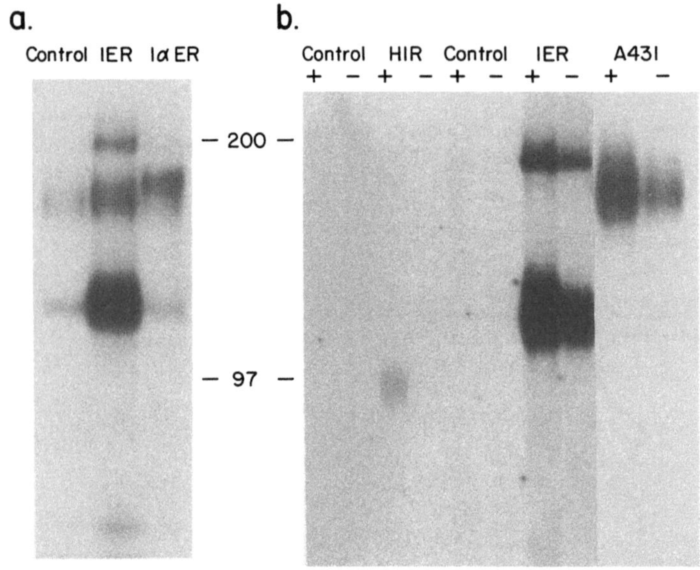

Fig. 9. Autophosphorylation of normal and chimeric receptors. COS-7 cells were transfected with the constructs indicated or with a control expression vector lacking a receptor cDNA insert. The EGF receptor was immunoprecipitated from A431 cells with MabRI. Detergent cell lysates were incubated with (+) insulin or EGF (A431 cells), or were mock-stimulated with PBS lacking hormone (−). Proteins were immunoprecipitated with human EGF receptor rabbit polyclonal antibody RK-2 (Kris et al. 1985) (a), human EGF receptor mouse monoclonal antibody RI (Waterfield et al. 1982) (b,A431), or human insulin receptor mouse monoclonal antibody CII25.3 (b)

the HIR construct and the amount of binding detected was proportional to the levels of the glycopolypeptides present in the transfected cells.

To test their intrinsic kinase activity, the expressed polypeptides were immunoprecipitated from lysates of transfected COS-7 cells using EGF receptor or insulin receptor antibodies. Insulin-stimulated autophosphorylation of the human insulin receptor β subunit was clearly detected above the endogenous COS-7 control, at substrate concentrations (100 μM ATP) required by the insulin receptor kinase (White et al. 1984; Petruzzelli et al. 1984) that result in reduced specific activity of the radioactive tracer (^{32}P-γ-ATP) and thus a weak signal. In contrast, only subpicomolar (0.1 pM) concentrations of ATP were needed to detect EGF-stimulated EGF-receptor kinase activity (Erneux et al. 1983), conditions under which the insulin receptor kinase is inactive.

As shown in Fig. 9a and b, the IER chimeric receptor displays insulin-stimulated autophosphorylation activity. Insulin stimulation of chimeric protein autophosphorylation was observed in immunoprecipitates obtained using antibodies to either the insulin receptor extracellular domain (Fig. 9b) or the EGF receptor intracellular domain (Fig. 9a). This experiment ruled out the possibility that small amounts of insulin-stimulated, endogenous COS cell insulin receptors catalyzed phosphorylation of the chimeric receptors in the immunoprecipitates. The properties of the chimeric IER kinase reflected the presence of the EGF receptor kinase domain because of the low ATP concentration (0.1 pM) required for autophosphorylation.

Interestingly, phosphorylation of the bands representing both the IβE fusion glycoprotein subunit (120,000 MW) and the uncleaved chimeric IER proreceptor (190,000 MW) were induced by insulin. Like the native EGF receptor (Erneux et al. 1983), maximal ligand induction was observed after 30 s at 4°C. Hence the EGF receptor tyrosine kinase retained its original enzymatic characteristics when controlled by the insulin binding domain within a chimeric receptor molecule.

By constructing a chimeric receptor molecule composed of extracellular insulin receptor sequences linked to EGF receptor transmembrane and cytoplasmic domains, we were able to demonstrate that binding of insulin to this IER chimeric receptor molecule induced tyrosine kinase activity in the EGF receptor-derived intracellular domain. The observed tyrosine kinase activity was characteristic of the EGF receptor in that it could be measured at subpicomolar ATP concentrations, conditions under which the phosphotransferase of the insulin receptor is inactive (White et al. 1984; Petruzzelli et al. 1984). Therefore, the IER chimeric receptor precursor must have been correctly processed and folded into a conformation that permitted communication between two entirely heterologous receptor domains, at the same time preserving the kinetic properties of the EGF receptor tyrosine kinase.

These results demonstrate directly that insulin and EGF receptors utilize a common mechanism for transmembrane signal transfer. This finding was unexpected since the primary structure and, indeed, the general architecture of the extracellular domains of these two proteins show little similarity. The only structural feature that is homologous between these receptors are regions with high cysteine content and conserved cysteine arrangement (Ullrich et al. 1985) that occur twice in the EGF receptor and once per insulin receptor α subunit.

How insulin and EGF receptors transduce a signal across the plasma membrane is not fully understood. It has been suggested that for both receptors the activation of the protein kinase is likely to be controlled by the state of receptor aggregation, which may be a consequence of ligand binding. What alternative mechanisms of transmembrane signal transfer exist for this receptor type? Domain communication through a transmembrane α helix of a monomeric molecule is difficult to envision, as would be interaction of the homologous, cysteine-rich units with a common, heterologous membrane component needed for signal transfer. We favor the notion that a ligand-induced conformational change in the receptor extracellular domain triggers receptor dimerization resulting in activation of intracellular enzymatic activities.

Whereas the EGF receptor may have to self-associate to induce kinase activity on the cytoplasmic side of the membrane, insulin and IGF-1 receptors (Massague and Czech 1982) that occur as disulfide-linked heterotetramers may be designed in this fashion to obviate the need for aggregation, and predispose these receptors for ligand responsiveness. The ability of the chimeric receptors to transduce a signal suggests that the structural differences between EGF and insulin receptors reflect only variations on a common mechanism for signal transfer. These experimental results open new avenues to investigate the mechanisms of transmembrane signalling and receptor function.

References

Alt FW, Kellems RD, Bertino JR, Schimke RT (1978) Selective multiplication of dihydrofolate reductase genes in methotrexate-resistant variants of cultured murine cells. J Biol Chem 253:1357-1370

Bargmann CI, Hung M-C, Weinberg RA (1986) The *neu* oncogene encodes an epidermal growth factor receptor-related protein. Nature 319:226-229

Carpenter G (1981) Epidermal growth factor receptor. Handbook exp. Pharmac. 57:89-132
Colbere-Garapin F, Horodniceanu F, Kourilsky P, Garapin A-C (1981) A new dominant hybrid selective marker for higher eukaryotic cells. J Mol Biol 150:1-14
Coussens L, Yang-Feng TL, Liao Y-C, Chen E, Gray A, McGrath J, Seeburg PH, Libermann TA, Schlessinger J, Francke U, Levinson A, Ullrich A (1985) Tyrosine kinase receptor with extensive homology to EGF receptor shares chromosomal location with *neu* oncogene. Science 230:1132-1139
Deutsch PJ, Wan CF, Rosen OM, Rubin CS (1983) Latent insulin receptors and possible receptor precursors in 3T3-L1 adipocytes. Proc Natl Acad Sci USA 80:133-136
Downward J, Yarden Y, Mayes E, Scrace G, Totty N, Stockwell P, Ullrich A, Schlessinger J, Waterfield MD (1984a) Close similarity of epidermal growth factor receptor and v-*erbB* oncogene protein sequences. Nature 307:521-527
Downward J, Parker P, Waterfield MD (1984b) Autophosphorylation sites on the epidermal growth factor receptor. Nature 311:483-485
Ebina Y, Ellis L, Jarnagin K, Edery M, Graf L, Clauser E, Ou J-H, Masiarz F, Kan YW, Goldfine ID, Rother RA, Rutter WJ (1985) The human insulin receptor cDNA: the structural basis for hormone-activated transmembrane signalling. Cell 40:747-758
Erneux C, Cohen S, Garbers D (1983) The kinetics of tyrosine phosphorylation by the purified epidermal growth factor receptor kinase of A-431 cells. J Biol Chem 258:4137-4142
Graham FL, Van der Eb AJ (1973) A new technique for the assay of infectivity of human adenovirus 5 DNA. Virology 52:456-467
Kahn CR (1985) The molecular mechanism of insulin action. Ann Rev Med 36:429-51
Kasuga M, Kahn CR, Hedo JA, Van Obberghen E, Yamada KM (1981) Insulin-induced receptor loss in cultured human lymphocytes is due to accelerated receptor degradation. Proc Natl Acad Sci USA 78:6917-6921
Kris RM, Lax I, Gullick W, Waterfield MD, Ullrich A, Fridkin M, Schlessinger J (1985) Antibodies against a synthetic peptide as a probe for the kinase activity of the avian EGF receptor and v-*erbB* protein. Cell 40:619-625
Massague J, Czech MP (1982) The subunit structures of two distinct receptors for insulin-like growth factors I and II and their relationship to the insulin receptor. J Biol Chem 257:5038-5045
Massague J, Pilch PF, Czech MP (1981) A unique proteolytic cleavage site on the β subunit of the insulin receptor. J Biol Chem 256:3182-3190
Neckameyer WS, Wang L-HW (1984) Molecular cloning and characterization of avian sarcoma virus UR2 and comparison of its transforming sequence with those of other avian sarcoma viruses. J Virol 50:914-921
Nilsen TW, Maroney PA, Goodwin RG, Rottman FM, Crittenden LB, Raines MA, Kung HJ (1985) C-erbB activation in ALV-induced erythroblastosis: Novel RNA processing and promoter insertion result in expression of an amino-truncated EGF receptor. Cell 41:719-726
Petruzzelli L, Herrera R, Rosen OM (1984) Insulin receptor is an insulin-dependent tyrosine protein kinase: Copurification of insulin-binding activity and protein kinase activity to homogeneity from human placenta. Proc Natl Acad Sci 81:3327-3331
Reed BC, Ronnett GV, Lane D (1981) Role of glycosylation and protein synthesis in insulin receptor metabolism by 3T3-L1 mouse adipocytes. Proc Natl Acad Sci USA 78:2908-2912
Sherr CJ, Rettenmier CW, Sacca R, Roussel MF, Look AT, Stanley ER (1985) The c-*fms* proto-oncogene product is related to the receptor for the mononuclear phagocyte growth factor CSF-1. Cell 41:665-676
Southern PJ, Berg P (1982) Transforming of mammalian cells to antibiotic resistance with a bacterial gene under control of the SV40 early region promoter. Journal of Molecular and Applied Genetics 1:327-341
Ullrich A, Coussens L, Hayflick JS, Dull TJ, Gray A, Tam AW, Lee J, Yarden Y, Libermann TA, Schlessinger J, Downward J, Mayes ELV, Whittle N, Waterfield MD, Seeburg PH (1984) Human epidermal growth factor receptor cDNA sequence and aberrant expression of the amplified gene in A431 epidermoid carcinoma cells. Nature 309: 418-425

Ullrich A, Bell JR, Chen EY, Herrera R, Petruzzelli LM, Dull TJ, Gray A, Coussens L, Liao Y-C, Tsubokawa M, Mason A, Seeburg PH, Grunfeld C, Rosen OM, Ramachandran J (1985) Human insulin receptor and its relationship to the tyrosine kinase family of oncogenes. Nature 313:756-761

Ullrich A, Gray A, Tam AW, Yang-Feng T, Tsubokawa M, Collins C, Henzel W, Le Bon T, Kathuria S, Chen E, Jacobs S, Francke U, Ramachandran J, Fujita-Yamaguchi Y (1986) Insulin-like growth factor I receptor primary structure: Comparison with insulin receptor suggests structural determinants that define functional specificity. EMBO J 5:2503-2512

Waterfield MD, Mayes ELV, Stroobant P, Bennet PLP, Young S, Goodfellow PN, Banting GS, Ozanne B (1982) A monoclonal antibody to the human epidermal growth factor receptor. J Cell Biochem 20:149-161

White MF, Haring H-U, Kasuga M, Kahn CR (1984) Kinetic properties and sites of autophosphorylation of the partially purified insulin receptor from hepatoma cells. J Biol Chem 259:255-264

Yamamoto T, Ikawa S, Akiyama T, Semba K, Nomura N, Miyajima N, Saito T, Toyoshima K (1986) Similarity of protein encoded by the human c-*erb*-B-2 gene to epidermal growth factor receptor. Nature 319:230-234

Yamamoto T, Nishida T, Miyajima N, Kawai S, Ooi T, Toyoshima K (1983) The *erb*-B gene of avian erythroblastosis virus is a member of the *src* gene family. Cell 35:71-78

Yarden Y, Schlessinger J (1985) The EGF receptor kinase: evidence for allosteric activation and intramolecular self-phosphorylation In: Evered D, Nugent J, Whelan J (eds) Growth factors in biology and medicine. Ciba Foundation Symposium 116, Pitman, London, pp 23-40

MIX
Papier aus verantwortungsvollen Quellen
Paper from responsible sources
FSC® C105338

If you have any concerns about our products,
you can contact us on
ProductSafety@springernature.com

In case Publisher is established outside the EU,
the EU authorized representative is:
Springer Nature Customer Service Center GmbH
Europaplatz 3, 69115 Heidelberg, Germany

Printed by Libri Plureos GmbH
in Hamburg, Germany